Science and Fictic

Science and Fiction – A Springer Series

This collection of entertaining and thought-provoking books will appeal equally to science buffs, scientists and science-fiction fans. It was born out of the recognition that scientific discovery and the creation of plausible fictional scenarios are often two sides of the same coin. Each relies on an understanding of the way the world works, coupled with the imaginative ability to invent new or alternative explanations—and even other worlds. Authored by practicing scientists as well as writers of hard science fiction, these books explore and exploit the borderlands between accepted science and its fictional counterpart. Uncovering mutual influences, promoting fruitful interaction, narrating and analyzing fictional scenarios, together they serve as a reaction vessel for inspired new ideas in science, technology, and beyond.

Whether fiction, fact, or forever undecidable: the Springer Series "Science and Fiction" intends to go where no one has gone before!

Its largely non-technical books take several different approaches. Journey with their authors as they

- Indulge in science speculation – describing intriguing, plausible yet unproven ideas;
- Exploit science fiction for educational purposes and as a means of promoting critical thinking;
- Explore the interplay of science and science fiction – throughout the history of the genre and looking ahead;
- Delve into related topics including, but not limited to: science as a creative process, the limits of science, interplay of literature and knowledge;

Readers can look forward to a broad range of topics, as intriguing as they are important. Here just a few by way of illustration:

- Time travel, superluminal travel, wormholes, teleportation
- Extraterrestrial intelligence and alien civilizations
- Artificial intelligence, planetary brains, the universe as a computer, virtual worlds
- Non-anthropocentric viewpoints
- Synthetic biology, genetic engineering, developing nanotechnologies
- Eco/infrastructure/meteorite-impact disaster scenarios
- Future scenarios, transhumanism, posthumanism, intelligence explosion
- Consciousness and mind manipulation

Kevin J. E. Walsh

Planets of the Known Galaxy

Fact and Fiction About the Nearest Stars and Their Worlds

🐴 Springer

Kevin J. E. Walsh
School of Geography, Earth and Atmospheric Sciences
University of Melbourne
Parkville, VIC, Australia

ISSN 2197-1188   ISSN 2197-1196   (electronic)
Science and Fiction
ISBN 978-3-031-68217-9   ISBN 978-3-031-68218-6   (eBook)
https://doi.org/10.1007/978-3-031-68218-6

This Springer imprint is published by the registered company Springer Nature Switzerland AG
The registered company address is: Gewerbestrasse 11, 6330 Cham, Switzerland

If disposing of this product, please recycle the paper.

Preface

This is not really a book about astronomy. Instead, it is about the environments of planets around other stars. There is a lot that we don't know, but as more observations are taken and more numerical simulations are performed, we can start to make some informed speculation about what these new planets are actually like. At the same time, science fiction writers have been imagining extrasolar planetary environments for decades. We have seen this process unfolding in the exploration of the planets of our own solar system. Since the 1950s, our perception of Mars has evolved from blurry images seen in Earth-based telescopes to panoramic vistas of rusty hills and valleys taken from the surface of the planet itself. The fictional depiction of Mars has developed from a world of canals and oases to a challenging Arctic desert. In short, Mars is no longer just a planet but has become a place. Some extrasolar planets have also started this journey.

I would like to thank the University of Melbourne School of Geography, Earth and Atmospheric Sciences, who were kind enough to host me during the time I was writing this book. Gemma van Hurk produced some nice illustrations that are included in the book. Thanks go to Trevor Quachri, editor of Analog Science Fiction and Fact, who recommended that I approach the editors of this book series with a proposal. Thanks also go to the team at Springer: Sujatha Chakkala, Cineha Dhakshinamoorthi, Lisa Scalone, Antje Endemann and Barbara Amorese. Above all, thanks go to my chief supporter and proofreader, my wife Giovanna.

Also, a disclaimer: there is a well-known planetary scientist named K.J. Walsh who works at the Southwest Research Institute in Boulder, Colorado. I cite his work a couple of times in this book, but he is not me.

Melbourne, VIC, Australia   Kevin J. E. Walsh
July 2024

Competing Interests The University of Melbourne has provided computing resources for this project. This project was undertaken with the assistance of resources and services from the National Computational Infrastructure (NCI), which is supported by the Australian Government. There are no competing interests.

Contents

About the Author

Kevin J. E. Walsh is a climate scientist and professorial fellow in the School of Geography, Earth and Atmospheric Sciences at the University of Melbourne, Australia. He has published over 100 peer-reviewed journal articles on topics ranging from climate to tropical meteorology to planetary science. He is also a regular contributor of popular science articles to Analog Science Fiction and Fact, and in 2023 was a finalist in their AnLab awards.

1

Introduction

The part of our Milky Way galaxy closest to Earth is a special place. It is the region of our galaxy that is best explored, the setting of hundreds of science fiction stories, and the part of the galaxy where it is most likely that life outside of our solar system will first be detected.

Life in other solar systems probably will not be found soon, but in the meantime the nearest part of our galaxy is full of interesting and even bizarre places. There are gloomy ocean worlds illuminated by the light of exploding stars, worlds where precious gems could be as common as pebbles, and planets eternally wandering between the stars like the Flying Dutchman. There are lava worlds, steam worlds, hot Jupiters, cold Jupiters and maybe even worlds like our Earth. We will call this place "the known galaxy".

Science fiction writers have invented many worlds in the known galaxy. From the balmy bays of Constance to the mountains of Rocannon's World and the deserts of Altair IV, the nearest stars have long been a setting for speculative fiction and film. Older tales and myths have naturally focused on the brightest stars, since the ancients did not know that some of these are nearby but others are very far away. But quite a few of the brightest stars are also within the known galaxy. The constellations that they form are the source of many legends, from the death of the centaur Chiron to the romantic tale of Qixi. All of this makes the known galaxy special.

The phrase "the known galaxy" has a particular meaning in science fiction, as it is the part of our galaxy that from the perspective of some imaginary year in the future would have already been settled and well explored by humans. Using this definition from the perspective of the year 2024, the known galaxy

© The Author(s), under exclusive license to Springer Nature Switzerland AG 2024
K. J. E. Walsh, *Planets of the Known Galaxy*, Science and Fiction,
https://doi.org/10.1007/978-3-031-68218-6_1

currently consists of Earth and its Moon. If we are a little more generous and include exploration by robot spacecraft, then much of our own solar system could now be described as "known". Some of these solar system worlds are starting to be viewed as places with their own personality, rather than just as dots in the sky. This transformation is already well under way for Mars: thanks to images from spacecraft and from generally realistic films like *The Martian* [1], we now have a mental image of the red planet as a cruel-looking, dusty, cold world with excellent scenery. In other words, we have a feeling for Mars as a place as well as a planet (see Fig. 1.1). One day, people will start to think like this about some of the better-known extrasolar planets. No spacecraft have explored the nearest stars, and they will not for a long time. Even so, we have learned a surprising amount about these extrasolar planets in the past couple of decades, from Earth- and space-based observations, from theory and from computer simulations.

So, for our purposes, how do we define the known galaxy? The known galaxy must be nearby, so we need to define quantitatively what is meant by "nearby". Distances to the stars are measured in a couple of different ways. The distance that light travels in a year is called a "light year" and is about 60,000 times the distance from the Earth to the Sun. By this measure, the nearest star group, the Alpha Centauri system, is about 4.3 light years away. Another measure of distance is the parsec, about 3.3 light years, and in this book we will use this as it is the distance unit that astronomers prefer. We also set the outer boundary of the known galaxy at the rather arbitrary distance of

Fig. 1.1 Martian landscape showing Aeolis Mons (also known as Mount Sharp), the central peak of the Gale crater, located in the equatorial region of the planet. Reprinted from https://photojournal.jpl.nasa.gov/. © 2015, NASA/JPL-Caltech/MSSS. Public domain

10 parsecs from Earth. Apart from being a nice round number, a distance of 10 parsecs has a couple of important aspects. A recent comprehensive survey [2] using spacecraft observations gives 540 known stars and other bodies closer than 10 parsecs, and this is a good starting point for our tour of the known galaxy. We will also talk briefly about some well-known stars and planets that are more distant, however.

Astronomers also know that, at a distance of 10 parsecs, the apparent brightness of a star (its 'magnitude') is defined to be the same as its actual (or 'absolute') magnitude. Stars closer than this appear brighter than their absolute magnitude, while stars further away appear dimmer. This explains why some perfectly ordinary but nearby stars appear bright in the night sky (like Alpha Centauri), while some enormous beacons of the galaxy, stars that are thousands of times brighter than Alpha Centauri, are hardly visible at all because they are too far away.

Of the 540 known objects closer than 10 parsecs, a minority are extrasolar planets. For some extrasolar planets, there is already an appreciation of them as places, through science fiction. Many of these stories are set on extrasolar planets that are Earth-like, albeit on worlds with some significant environmental differences from our own world. The holy grail of the exploration of the known galaxy is to find another planet like Earth. This discovery would confirm that there are worlds out there that could potentially host life, in other words, that are "habitable". Actual detection of life itself might also be possible in the future through advanced astronomical techniques [3, 4]. Discovery of life on another planet would mean that we humans are not alone in the known galaxy. Well, actually, if you ask astronomers, almost all of them say that such a discovery is just about inevitable, although it will likely take some decades. It will still be very big news when it happens, though.

An Earth-like world is by definition habitable, but what does "habitable" really mean? This isn't such a dumb question when you think about it. Even on Earth, there are varying degrees of habitability: for example, the Hawaiian Islands are rather more habitable than the South Pole. This concept can be extended to the solar system. Despite Mars being the most Earth-like planet, the Earth's South Pole is far more habitable than any place on Mars. It is easy to understand why. For a start, in Antarctica, you don't need a space suit to survive on the surface, but you do on Mars. Also, in Antarctica there are tons of water in the form of surface snow and ice easily available to be used. On Mars, water is available but is harder to find as much of it is underground. But any place on Mars is more habitable than any place on the surface of Venus, where because of the baking heat and crushing pressures, in order to stay alive each astronaut would be required to wear an uncomfortably large, fiercely heat-shielded armored mini-submarine. Also, when we say "habitable", do we mean habitable for human beings? Or only for bacteria? A planet could be bursting with life and have a perfect climate with mild temperatures in many locations but might have an atmosphere that humans cannot breathe, like the planet Pandora in the movie *Avatar* [5, 6]. Does the presence of life there make the planet "habitable"?

In this book, we make a broad definition: "habitable" means that a planet could potentially host life of some kind [7]. Meteorologically, we will assume that this means that the planet could possibly have liquid water on its surface. Water is an important component of terrestrial life and is crucial for human habitability. Still, other liquids might form the basis of exotic extraterrestrial life: liquid ammonia has been mentioned quite frequently, both in the scientific literature and in science fiction [8]. But water has some very significant advantages. It remains a liquid over a reasonably large range of temperatures at the atmospheric pressures congenial to human beings, it does not react with an oxygen atmosphere, it is abundant, and so on. Typically, this range of temperatures is used to define a "habitable zone" [9], a region around a star where the planet is not too close or too far from its star, so that the amount of stellar radiation received is not too much or too little, and so that a planet in the habitable zone is not too hot or too cold.

Now of course in this early stage of extrasolar planet exploration, there remain varying definitions of the inner and outer limits of the habitable zone, depending on the assumed relationship between incoming stellar radiation and the actual surface temperature of the planet [10, 11]. This is important because while the incoming stellar radiation amount for a chosen planet is usually known accurately, the surface temperature isn't. This is because the atmospheric composition can greatly change the surface temperature through the greenhouse effect (see below). Even in our own solar system, where both stellar radiation and surface temperature values are known reasonably well, estimates of the inner and outer limits of the habitable zone vary. In fact, some calculations put the inner limit of the habitable zone uncomfortably close to the actual current orbit of the Earth [12]. Putting that scary thought aside for the moment, we will denote planets that orbit within a plausible definition of the habitable zone as "habitable zone planets".

Note that "habitable zone planets" might not necessarily be very comfortable places for human beings. Many planets within the habitable zone will not be habitable at all, being for various reasons either too hot or too cold. While many of these worlds will have liquid water on their surfaces, it could be in the form of oceans hot enough to steam up a sauna. Quite a few will be ocean planets with no land surface. Many, perhaps the overwhelming majority, will have atmospheres that have no oxygen or not enough oxygen for humans to breathe. Some, a few, might have decent temperatures and a breathable atmosphere but absurdly high levels of planetary volcanism or meteor strike, or crushingly high surface gravity. There are many possible ways in which a habitable zone planet cannot be truly Earth-like, that is, with temperatures that are comfortable for human beings somewhere on the land surface of the planet, and with an atmosphere that is breathable by human beings indefinitely. The class of Earth-like planets is a very restricted subset of all planets, and it currently has a membership of one (i.e. Earth). We will likely find others in the future but given how easily things can go wrong in the evolution of the climate of a planet, I suspect that genuinely Earth-like planets will not be very common.

For extrasolar planets, we currently do not have all of the information required to determine for certain whether a particular planet has either habitable zone or Earth-like levels of habitability. On the other hand, it is possible to say whether it is likely out of the running or not. So far, astronomers have been able to determine the size of some extrasolar planets as well as their densities, along with some details regarding the atmospheric composition of a few larger worlds, as well as the amount of stellar radiation received. This basic information has been used for a number of attempts to estimate the potential habitability of the various extrasolar worlds, as a way of prioritizing their exploration.

Potential habitability has been measured by habitability scores or indices, based on a few assumptions of the likely links between habitability and known planetary characteristics. A relatively simple index, the Earth Similarity Index (ESI) [13], is based only on the planetary size and the incoming stellar radiation. Based on the ESI, as of 2024 the most similar extrasolar planet to Earth is Teegarden's Star b [14] (see Chap. 7; for an overview of how stars and planets are named, see the end of this chapter). This index is not directly related to how hospitable the planet is to life, as the ESI includes no direct information about the actual surface conditions. Nevertheless, the ESI will be partially related to habitability, as planets like Jupiter receive rather low ESI scores while planets like Mars receive much higher scores because they are much more Earth-like.

There are now a number of these indices, and each has their advantages and disadvantages [15]. A recent attempt to estimate the actual surface climate is the so-called "Statistical-likelihood Exo-Planetary Habitability Index" or SEPHI [16]. In this method, a combination of theory and observations from our own solar system are used to define four criteria that determine potential habitability. Planets with large amounts of gas would have crushing atmospheres that would make them very inhospitable to human beings or most other terrestrial life. Thus an estimate of the gas content of a planet is used as the first habitability criterion. Even mostly rocky worlds, though, can have atmospheres that are too thick, so a second criterion is imposed to estimate whether the atmosphere is thin enough for the planet to be habitable. A planet also needs mild temperatures that ensure the presence of liquid water, as this is likely crucial for life, so this is the third criterion. Finally, planets like Earth have a magnetic field generated by the rotation of an iron core. A planet either with slow rotation or little iron will likely have a weak magnetic field. The Earth's magnetic field shields its outer atmosphere from the relentless solar wind, a stream of electrons, protons and other particles emanating from the Sun that eventually could strip the Earth's atmosphere away if the magnetic field were too weak. Thus an estimate of the strength of its planetary magnetic field is the final habitability criterion in the SEPHI index. We will use information from this index to describe the possible environments of individual extrasolar planets.

The stars and planets of the known galaxy consist of a number of different types. The star types will be described first because they can have a strong influence on the habitability of their planets.

Star Types

Stars evolve with age [17]. They shine through the process of hydrogen fusion, a forced merger of hydrogen atoms that releases a tremendous amount of nuclear energy. Those that are still burning their hydrogen fuel are called main sequence stars, while stars that have exhausted almost all of their hydrogen and have expanded are called giant or supergiant stars. As these stars grow bigger, they become brighter and much redder. Then, often after a disastrous explosion or two, stars the size of our Sun leave behind a hot dense remnant known as a white dwarf. Stars substantially larger than our Sun sometimes end up as even denser neutron stars, and some wind up as black holes with such strong gravity that not even light can escape from them. For stars like the Sun, this whole process will take about 10 billion years; for very large stars, it will take only millions of years; but for small stars, it will be trillions.

Classification by Temperature Here we arrange stars in order of temperature, from hot to cold [18]. Hot stars are mostly blue and cold stars mostly red. To quantify star and planetary temperatures, we will generally use degrees Kelvin, a temperature scale that is 0 at the lowest possible temperature where molecular motion ceases, known as absolute zero. The freezing point of fresh water at sea level on Earth is 273.16 K, and the interval between degrees in the Kelvin scale is the same as in the Celsius temperature scale (°C). In degrees C, the freezing point of water is 0 °C and so in the Celsius scale, absolute zero is minus 273.16 °C. We will occasionally use degrees Celsius to make a point about planets with Earth-like temperatures, and we will sometimes also use degrees Fahrenheit for the same reason.

O Stars This very rare type has surface temperatures of more than 25,000 K and emits radiation mostly in the ultraviolet part of the spectrum. These large stars typically spend less than 10 million years on the main sequence. The nearest one is Zeta Ophiuchi, quite distant at about 120 parsecs away [19], in other words well outside our definition of the boundary of the known galaxy. It has about 20 times the Sun's mass and 74,000 times its brightness [20]. Between Zeta Ophiuchi and our own solar system, there is so much interstellar dust blocking this star's light that if the dust were removed, Zeta Ophiuchi would be one of the brightest stars in our night sky [21].

B Stars There are many more B stars than O stars but they are still relatively rare, comprising only about 0.1% of stars in the Sun's neighborhood [22]. They are typically very blue and short lived, although not quite as short lived as O stars. The nearest one is a component of the multi-star system that comprises Regulus, the brightest star in the zodiacal constellation Leo, about 24 parsecs away [19].

A Stars These white stars are much more numerous than O or B stars and the classic example is Sirius A, the brighter of the two stars that make up the Sirius system (the other star, a white dwarf, is naturally designated Sirius B). Sirius is the brightest star in the night sky due to its proximity to Earth, being only 2.6 parsecs away. It is a few hundred million years old and about 25 times brighter than the Sun [23] (see Chap. 5).

G Stars The Sun is a yellow G star and is of course the closest example of this type, although Alpha Centauri A is the second closest, at only 1.3 parsecs away (see Chap. 2). The Sun is about 4.6 billion years old and still has several billion years to go before it starts turning into a red giant [24]. Other nearby, older G stars appear to have already started this process (e.g. 61 Virginis; see Chap. 2).

K Stars Somewhat redder and somewhat dimmer than G stars, these long-lived, generally well-behaved stars are prime candidates in searches for habitable planets. The nearest one is Alpha Centauri B (see Chap. 2).

M Stars These common-as-dirt red stars are of keen interest because they are so numerous and because so many planets have already been discovered orbiting them. They are very long lived but often their brightness is not very stable, particularly when they are younger. This may make long-term habitability of their planets more difficult. The nearest one, and the nearest known star, is Proxima Centauri (see Chap. 2), part of the Alpha Centauri stellar system.

There are also smaller, dimmer types of luminous bodies that bridge the gap between stars and planets. Brown dwarfs [25, 26], as these smaller bodies are known, are too small to shine by hydrogen fusion, but instead the larger ones fuse either deuterium (a heavier version of hydrogen) or lithium. These molecules fuse at a lower temperature than hydrogen, which is why experimental fusion reactors on Earth use deuterium and not hydrogen. Brown dwarfs are of interest here mainly because they may act as spoilers in some planetary systems, with their gravity making planetary formation around some stars more difficult. They come in a number of different types:

L Dwarfs Since this classification is color- and temperature-based rather than mass-based, this class consists of a mix of stars and brown dwarfs. That is, it is possible for both a small star and a big brown dwarf to have similar temperatures and redness, and thus the same classification (there are also some M-class brown dwarfs). These L dwarfs have clouds in their upper atmospheres comprised of condensed metal or rock droplets [27]. The nearest one is Luhman 16A, part of the third-nearest stellar system, about 2 parsecs away (Chap. 5) [28].

T Dwarfs T dwarfs are cooler and so metal clouds solidify and sink lower into the hotter parts of the atmosphere, leaving many T dwarfs with fewer clouds than L dwarfs [27]. The nearest example known is Luhman 16B, in the same system as the L dwarf Luhman 16A.

Y Dwarfs These dwarfs are so small that they do not burn any fuel at all and only warm themselves by slow contraction, as does Jupiter [29]. They would often also look rather like Jupiter, with various horizontal cloud bands of water and ammonia. The nearest one known has the catchy name WISE 0855−0714 and at a distance of about 2.2 parsecs is the fourth-closest known stellar system to the Sun [30]. Its cloud tops are quite cold, with an estimated temperature of only about −30 °C (−22 °F), and it has a mass only about 3–10 times that of Jupiter.

There is a hoary old mnemonic that is used to remember the order of stellar types: "Oh Be A Fine Girl/Guy, Kiss Me." To which could be added any number of phrases from the Internet for the more recent brown dwarf classes "L, T and Y". *Caveat lector.*

Planetary Types in the Solar System and Beyond

Stellar classification is based on observations of millions of stars and on well-established theoretical calculations of their behaviour and evolution. In contrast, the classification of planetary types is in its infancy and as a result there is no generally accepted classification scheme yet. Here we will use a simplified version of the scheme recently proposed by Ravi Kopparapu of NASA [31], with some of our own extensions to include important members of our own solar system. This scheme classifies planets on the basis of their radius and temperature. Planets with small radii are likely to be comprised mostly of rocky elements, as Earth is. This is because there is a lot more hydrogen and helium gas in the Universe than there are rocky elements like silicon and iron, and so the only way for a planet to become as large as Jupiter is to collect a lot of gas rather than rocks. Thus as planetary radius increases, the percentage of the planet's mass made up of gas tends to increase as well. As a result, while planets like Earth are mostly rocky, large planets like Jupiter are mostly gas.

The other factor in the Kopparapu classification scheme, temperature, is of obvious importance for planetary evolution in our solar system. Earth and Venus have very similar masses but very different atmospheres and surface temperatures. This is largely blamed on the fact that Venus receives about twice the solar radiation that Earth does, and this has strongly affected the development of its atmosphere [32]. The three broad temperature categories that are used in the Kopparapu scheme are "hot", "warm" and "cold", and the boundaries between these categories are set on the basis of whether certain molecules can condense into clouds in the atmosphere of a planet or not. "Hot" planets are ones where zinc sulphide clouds can form but not water clouds. Why zinc sulphide? It is a metal that is likely to form clouds on hot extrasolar planets ("exoplanets") and some work has been done on that. Some other types of metal clouds would still be present at even higher temperatures on some planets, like clouds of molten iron droplets for instance, because iron condenses to form cloud droplets at a considerably higher temperature than zinc sulphide. In an extension of the Kopparapu classification scheme, we'll call these planets "very hot", as they are typically too hot for zinc sulphide clouds to form. "Warm" planets are where water clouds can form (like Earth), while "cold" planets are where carbon dioxide clouds can form, but methane clouds cannot. For colder planets where methane clouds can form, we will designate these as "very cold". Even so, this temperature scheme is a general description of the planetary environment and does not necessarily apply to every part of the planet being classified. For instance, there is water ice in the polar regions of Mercury, despite the planet being well inside the "hot" classification zone [33]. On Mars, carbon dioxide clouds can form during the winter at the poles and also at high altitudes, despite this planet being in the "warm" category [34]. Like all general classifications, the scheme misses some of the details of actual planetary conditions.

Table 1.1 shows the classification that is used here. Planetary sizes are given on the horizontal axis, with size ranges given in multiples of Earth radii. Since brown dwarfs are about the same size as super-Jovians but more massive, a mass criterion is used to define them instead, measured in multiples of the mass of Jupiter (MJ) [35]. On the vertical axis, incoming stellar radiation is given, defined as a multiple of Earth values. There are a lot of blank boxes in Table 1.1 because there are many planetary types that do not exist in our own solar system but are found in other solar systems. In this table, we have the following planetary categories in order of descending mass. For extrasolar worlds the smaller planets are very likely to be more numerous than the larger ones, just as they are in our Solar System:

Brown Dwarfs (see above for a discussion of the different types of brown dwarfs).

Super-Jovians (More Than 14.3 Times the Radius of Earth) This category overlaps with Y class brown dwarfs (see above).

Jovians (Between 6 and 14.3 Times Earth Radius) Typical examples are Jupiter and Saturn from our solar system. These are both "cold Jovians" in the classification, although there are numerous examples of warm and hot Jovians that have been discovered in other stellar systems. Jovian planets are mostly comprised of hydrogen and helium gas, although Jupiter and other Jovians are believed also to have a substantial core of rocky elements [36].

Sub-Jovians (Between 3.5 and 6.0 Times Earth Radius) Uranus and Neptune are examples of very cold sub-Jovians. Sub-Jovians, sometimes called "ice giants", are planets that still have large atmospheres but their composition is not dominated by hydrogen and helium, as the majority of their mass consists of rock and ice. There is still considerable uncertainty about their interior structure. For Uranus, the "ice" component may actually be a large, hot fluid of water, ammonia and methane, as when astronomers say "ice", they mean

Table 1.1 Planetary types with examples from our Solar System, using a simplified version of the classification of Kopparapu et al. [31] but with additional categories. For simplicity, the radiation categories defined by Kopparapu et al. for rocky worlds are here used for all bodies, as these categories vary with mass

Radiation (Earth = 1.0)	Radius (Earth = 1.0)								
	Sub-rocky (0.1–0.5)	Rocky (0.5–0.8)	Earth-size (0.8–1.25)	Super-Earth (1.25–1.75)	Sub-Neptune (1.75–3.5)	Sub-Jovian (3.5–6.0)	Jovian (6.0–14.3)	Super-Jovian (>14.3 Earth radii; >1 MJ)	Brown dwarf (>13 MJ)
Very Cold (<0.0035)	Pluto (and many others)			Planet Nine?					
Cold (0.0035–0.28)	Titan (and many others)				Uranus, Neptune		Jupiter, Saturn		
Warm (0.28–1)	Moon	Mars	Earth						
Hot (1–182)	Mercury		Venus						
Very Hot (>182)									

either the liquid or solid form of compounds with melting points that are relatively low [37]. Above Uranus's global "ocean", there is a transition to a dense atmosphere of hydrogen and helium, with some small amounts of methane and other gases.

Sub-Neptunes (Between 1.75 and 3.5 Times Earth Radius) There aren't any in our solar system but elsewhere in our galaxy many have been discovered [38]. Superficially, they are expected to be gassy planets like the sub-Jovians but smaller. Some of them may be "gas dwarfs", with a rocky core surrounded by an envelope of hydrogen and helium gas. Others may contain a substantial amount of water [39]. There is however another type of planet in this category, the so-called "Mega-Earths", massive rocky planets without the substantial atmosphere of most sub-Neptunes [40]. One possible way for such a planet to originate would be if a Jovian were orbiting so close to its star that its atmosphere was gradually stripped away, leaving behind a hot, rocky core [41]. These are known as chthonian ("of the earth") planets, named after the Greek gods of the underworld [42]. No confirmed examples of this type have been found but it appears that there are some very hot extrasolar Jovians that are on their way to becoming chthonian, with their atmospheres being blasted into space by the intense radiation of their nearby stars. One distant recently discovered world, TOI-849 b, may already be chthonian but this is not yet confirmed [43].

Super-Earths (Between 1.25 and 1.75 Times Earth Radius) Another category that has no known examples in our solar system, unless we count the as-yet-unconfirmed Planet Nine (see later in this chapter). These are planets that are mostly rock or ice, with only small atmospheres that do not constitute a substantial fraction of the planetary mass, although such an atmosphere could still be very thick compared to Earth's. Many of these planets will be water worlds with no solid land surface, due to the ability of their gravity to attract watery asteroids and comets during their formation [44]. The amount of water on these planets is predicted to be very variable, though, with some of them being extreme desert worlds.

Earth-Size (Between 0.8 and 1.25 Times Earth Radius) [45] The title is self-explanatory, but even though these planets are comparable in size to Earth, their characteristics will vary widely. Even within our own solar system, both Venus and Earth are Earth-size but have very different environments. Like super-Earths, their atmospheres are classified as thin, although like Venus, many planets could have atmospheres with surface pressures many times that of Earth.

Rocky (Between 0.5 and 0.8 Times Earth Radius) Rocky planets have small atmospheres like super-Earths and Earth-size worlds but are less likely to be water worlds and more likely to eventually lose the water that they started with [46]. An example in our solar system of a rocky planet that has lost most of its original water is Mars.

Sub-rocky (Between 0.1 and 0.5 Times Earth Radius) The lower boundary of this category is set approximately at the minimum radius that a planet would need to still have enough of its own gravity to shape itself into a sphere, as opposed to the irregular shape of many smaller asteroids and comets [47]. Worlds in this category that are in the warm or hot zone are likely to have little water, having either lost it or never had it, and would also be warm enough for most gases to escape their weak gravity, leaving behind a thin or negligible atmosphere. Examples in our solar system include Mercury and the Moon. Sub-rocky worlds in the cold zone are more likely to be cold enough to retain a substantial atmosphere, like Saturn's moon Titan, for instance [48]. A few extrasolar sub-rocky planets have been found but because of their small size they are difficult to detect, despite likely being very numerous [49].

Various techniques have been used to discover extrasolar planets, but quite a few readers may already be familiar with these techniques and how they work. In any event, the history and methods of extrasolar planet discovery are not the primary focus of this book, so I've put a brief description of these techniques at the end of this chapter.

Distances and Temperatures in Stellar Planetary Systems

Typically, distances from stars to their planets are measured in multiples of the distance from the Earth to the Sun. This distance, about 140 million kilometers, is called an astronomical unit, or a.u. for short, and we will use this for giving distances within stellar systems. For temperature, the vast majority of extrasolar planetary temperatures have not yet been measured directly. Instead, we will give their equilibrium temperatures, a calculated average temperature for the planet that assumes a balance between the amount of stellar radiation absorbed by the planet and the amount of infrared radiation or heat that is emitted by it. The closer a planet is to a particular star, the higher the amount of radiation that it absorbs and the higher its equilibrium temperature. The equilibrium temperature also depends upon the fraction of stellar radiation reflected by the planet, known as the albedo. If the albedo were higher, like that of fresh snow for instance, more stellar radiation would be reflected back into space and the calculated equilibrium temperature would be lower. For easy comparison to Earth, we will assume that the albedos of the extrasolar planets are the same as Earth's unless we have observational evidence to the contrary. A slight wrinkle in this assumption is that M stars radiate mostly in the infrared rather than visible light like our Sun. Infrared radiation does not reflect well, so instead it would be largely absorbed by any planets orbiting M stars. Thus for M stars, the effective albedo is likely to be closer to zero, and for these stars we will also give an equilibrium temperature calculated using an albedo of zero.

*   *   *

We start our tour of the known galaxy with a planet that may not even exist.

In our solar system, just beyond the orbit of Neptune or about 30–50 a.u. from the Sun, there are a large number of sub-rocky and smaller objects, often with oval-shaped (known as "eccentric") orbits. These are known as Kuiper belt objects (KBOs—Kuiper is pronounced to rhyme with "viper") [50]. This belt is thought to consist of material left over from the formation of the solar system. During formation, in the inner part of the solar system near the Sun, gas and dust were collected by the Sun's gravity. In that part of the solar system, the density of these KBO-sized objects was a lot higher and so they eventually clumped together into planets. In the Kuiper belt, being so much further from the dusty central region, the KBOs failed to aggregate into planets and instead formed a large number of smaller worlds [51]. More than 100,000 KBOs larger than 100 km in diameter are likely to exist [52]. The largest known KBO is Triton, the biggest moon of Neptune, believed to have been captured by Neptune billions of years ago. Pluto is also a KBO (no, I'm not going to discuss whether Pluto should be reinstated as a planet). There are other objects that are usually further away from the Sun than the Kuiper belt but often pass through it. These belong to the "scattered disk", so called because it owes its likely origin to gravitational scattering of small bodies by the giant planets, particularly Neptune. Even further out is the Oort cloud, a halo of comets at least 2000 a.u. from the Sun and extending halfway to the nearest star.

The mechanisms explaining the formation of the KBOs are still being debated, but a further problem arose in 2004 with the discovery of sub-rocky world named Sedna [53]. This very small, very red world has a highly eccentric orbit ranging in distance between 76 and about 900 a.u from the Sun, or outside the Kuiper belt and traversing the scattered disk region. Sedna's strange orbit cannot be easily explained by gravitational interactions with Neptune, as Sedna is too far away. The discovery in 2014 of another object with a Sedna-like orbit deepened the mystery [54]. This led some astronomers to propose that the only explanation was the presence of another undiscovered planet somewhere outside the Kuiper belt whose gravity was forcing Sedna and other objects into their strange orbits. Known for the time being as Planet Nine, this planet was initially proposed to be a super-Earth with an average distance from the Sun of about 380 a.u, although now there are various estimates of this distance [55].

A super-Earth at 380 a.u. from the Sun would receive only a negligible amount of solar radiation, with a resulting equilibrium temperature of only about 13 K, or very near the melting point of hydrogen. But it would be a mistake to assume that this planet would necessarily have such a low surface temperature. In 1999, David Stevenson wrote a prescient article in the prestigious journal Nature, pointing out that even in the far, cold reaches of interstellar space, largish planets would still release internal heat left over from their formation and from the slow decay of radioactive elements inside them [56]. This would likely ensure a much higher surface temperature and a much more interesting planetary environment as a result. This process also occurs on Earth and is the reason that the inner core of our planet consists of molten rock with a typical temperature of about 5600 K [57]. On Earth, not much of this internal heat reaches the surface, only about 1/3000th of the amount of visible radiation arriving from the Sun [58]. A super-Earth could have considerably more internal heat reaching its surface than Earth because it would be a larger world with the potential to contain more radioactive elements. If such a planet also had a thick hydrogen atmosphere, it would have a substantial greenhouse effect and much warmer surface temperatures.

The greenhouse effect gets a lot of press, but mostly this is about what is known as the "enhanced" greenhouse effect, the extra warming caused by the increased carbon dioxide put into the atmosphere by the burning of fossil fuels since the beginning of the Industrial Revolution in the late eighteenth century. In planetary science, the greenhouse effect instead refers to the so-called "natural" greenhouse effect, the warming provided by the water vapor, carbon dioxide and other greenhouse gases that have been in the Earth's atmosphere in varying amounts for billions of years. Ironically, the greenhouse effect doesn't work like a greenhouse. A glass greenhouse is warmer than the outside air because it lets sunlight in but the glass walls stop winds from taking away the extra heat. In contrast, the planetary greenhouse works like a blanket: sunlight heats up the surface, so heat is emitted by the surface but the greenhouse gases absorb some of the heat radiation and emit it back towards the surface instead of out to space. Almost all planets with atmospheres will have some greenhouse effect. On Earth, the natural greenhouse effect is what makes Earth's average surface temperature about 33 °C (59 °F) warmer than its equilibrium temperature of 255 K (−18 °C; 0 °F) [59]. Mars does have a greenhouse effect but it only gives a warming of a few degrees above its equilibrium temperature because the atmosphere of Mars is very thin, despite being mostly comprised of the greenhouse gas carbon dioxide. Venus has a thick atmosphere comprised mostly of carbon dioxide and thus has an enormous greenhouse effect of several hundred degrees, which largely accounts for its hot, grossly uninhabitable surface.

Venus is likely hot because of the so-called "runaway greenhouse" effect [60]. The Sun's brightness has increased as it has become older. As temperatures started to rise on Venus, so did the evaporation of its water. Now water vapor is a greenhouse gas and so it is an effective absorber of infrared radiation. As the atmospheric temperature increased, at some point the amount of water vapor in the atmosphere became so large that the infrared radiation emitted by the hot surface was trapped by the water vapor and so was no longer able to escape effectively to space to cool the planet down. After that, the surface temperature needed to continue to rise until so much infrared radiation was being radiated to space by hot, non-greenhouse gases in the atmosphere that the radiation emitted to space finally became large enough to compensate for the greenhouse trapping effect of the water vapor. At this new, much higher temperature, the oceans boiled away and most of the water vapor then escaped into space, thus completing the runaway greenhouse process. Whether a planet receives enough stellar radiation to undergo this effect is an important potential limit on its habitability.

The inner limit of the habitable zone is typically set at the distance where runaway greenhouse occurs. There is also a related concept known as the "moist greenhouse", where the planet is not yet undergoing runaway greenhouse but is so hot that water vapor becomes a major constituent of the atmosphere, and the planet starts losing a lot of water to space. This takes place at a slightly lower temperatures than the runaway greenhouse, and it is thought that Venus initially took this route [61]. Also, while Venus has no oceans today, there is some evidence that it might have had liquid water on its surface as recently as a billion years ago [62]. This gives an inner boundary of the habitable zone a little closer to the star than the runaway greenhouse distance, known as the "early Venus" limit. The outer boundary of the habitable zone is typically set by the "maximum greenhouse" limit, where reflection of incoming light by carbon dioxide molecules starts to outweigh the warming effect of this greenhouse gas. There is also an "early Mars" limit a little further out, based on geological evidence that Mars appeared to have had liquid water on its surface about 3.8 billion years ago, and so was in the habitable zone at that time [63]. The region between the early Venus and early Mars limits is known as the optimistic habitable zone, while the smaller region between the runaway and maximum greenhouse limits is known as the conservative habitable zone.

The greenhouse effect can even have a big impact on the surface climate of worlds that receive almost no incoming stellar radiation. In the far reaches of the outer solar system, a super-Earth with a reasonable amount of internal heat and a hydrogen atmosphere with a surface atmospheric pressure about 1000 times that of Earth could have a surface temperature above freezing and might even have a water ocean [56]. We'll call this hypothetical planet Stevenson's World. Planet Nine could be a Stevenson's World but it has not been found yet, and due to its distance and dimness will be very difficult to spot. In fact, it is even possible that it could be a small black hole, which would make it even more difficult but still not impossible to find [64]. There are also some other possible explanations for the orbits of Sedna and similar objects in the outer solar system, so it is not clear yet whether Planet Nine even exists [65]. Nevertheless, we know that there is at least one extrasolar planet in a similarly distant orbit from its host star. The planet HD 106906 b, a super-Jovian about 103 parsecs away, has a current separation from its star of about 740 a.u or about 25 times the distance from the Sun to Neptune. The most likely explanation for this unusually large distance is that HD 106906 b was kicked out of its stellar system, but then instead of escaping entirely was nudged back into its current orbit by the gravity of a passing star [66].

It is almost certain that many other worlds like Planet Nine are out there floating between the stars. Simulations of planetary formation show that planets are routinely ejected from stellar systems during the formation process and end up wandering through interstellar space, becoming what are known as "free-floating planets" or "rogue planets" [67]. This is not a new idea, as astronomers and science fiction writers have been speculating about rogue planets for decades. Estimates of the number of such worlds are very uncertain, but for worlds of sub-rocky size or greater, it is conceivable that there could be dozens or hundreds of times as many of them as there are stars [68–70]. There are also some super-Jovians that have already been identified as rogue planets, although many of these are big enough to have formed by themselves in the same way that stars form, rather than having been ejected from a stellar system; the nearby Y dwarf WISE 0855−0714 is an example (see Chap. 5). Nor is the ejection of planets limited to those other mysterious stellar systems somewhere out there in the universe. Simulations of the formation of our own solar system indicate that one way for our system to end up with its current rather unusual arrangement of planets is for two things to have occurred: the giant planets Jupiter and Saturn must have wandered around the solar system before settling down into their current orbits, and in the process at least one planet about the size of Uranus was ejected [71].

This poses the question: if this planet were ejected, where is it now? The answer is that no one knows, and in fact it is likely that no one will ever know. We can estimate the probability of various scenarios of planetary formation but the process of planetary ejection is random and violent, and so the precise details are difficult or impossible to reconstruct after the event. It would be a bit like trying to figure out what a sandcastle looked like before the kids thoroughly stomped on it. But somewhere out there, likely still wandering between the stars, is the solar system's long-lost planet, which for our purposes we will call Chaos. Like Chaos, some rogue planets could be sub-Jovians, but they could also be very diverse. Smaller, rocky ones could be quite dry if they were ejected from the inner, hotter part of a stellar system, where planets have difficulty keeping liquid water on their surfaces. These rogue planets would end up as extremely cold desert worlds, a kind of frozen Mercury with a surface lit only by starlight. More massive worlds, or ones ejected from the colder outer part of a solar system, would be more likely to retain an atmosphere initially, although for most of them once the planet was in interstellar space the atmosphere would freeze out on the surface. Such planets might have oceans of liquid hydrogen rather than water, and Chaos could be one such world. Stevenson's World is also another possible planet type, but it needs a dense hydrogen atmosphere to generate a large greenhouse effect. These planets would probably be larger and therefore rarer than smaller, rocky rogue planets.

Of course, there is another possible fate for long-lost Chaos: it could actually be Planet Nine itself [55].

Yet another type of rogue planet has been suggested by Dorian Abbot and Eric Switzer of the University of Chicago. In an article published in 2011 [72], they outlined what they called a Steppenwolf planet, so called because it wanders the interstellar galactic steppe like a lone wolf, a bit like Harry Haller, the protagonist in Hermann Hesse's eponymous novel [73]. In their paper, Abbot and Switzer asked whether a planet in interstellar space could retain a subsurface water ocean if it were protected by a thick layer of ice. This is a reasonable question because there are at least a couple of sub-rocky worlds in our own solar system that are like this, namely Enceladus, one of the moons of Saturn, and Europa, one of the moons of Jupiter. Enceladus has an ice layer about 25 km thick overlaying a 30 km deep ocean [74], while Europa's ice layer is possibly thinner, at 10 km or so, also with a deep ocean [75]. The icy surface of these worlds is so cold that the ice is as hard as rock, but the liquid subsurface oceans are warmed up by what is known as tidal heating. This heating occurs on our Earth as well: the Moon pulls on the Earth's solid and liquid surface, moving it back and forth a little as the Moon orbits the Earth, and the resulting flexing and friction causes a small amount of heating. On Earth, this is tiny compared with other heat sources, though [76]. Much larger is the tidal heating caused by Saturn and Jupiter on some of their moons. Jupiter's moon Io is so strongly tidally heated that it has a mostly molten surface and large active volcanoes (see Fig. 1.2) [77]. Europa is further away from Jupiter than Io so the tidal heating on Europa is less, but it is still likely a major contributing factor to the maintenance of Europa's subsurface ocean [78].

For a Steppenwolf, there will be no external tidal heating on one that is wandering around by itself, so its required heat source has to be internal heat. Based on typical internal heating rates, Abbot and Switzer's calculations estimate that if a Steppenwolf had about the same amount of water as Earth does, it would need to be a super-Earth with about 3.5 times the mass of Earth in order to retain a subsurface ocean underneath an ice layer several kilometers thick. If a Steppenwolf had ten times the amount of water Earth does, which is a realistic scenario for some worlds, it would only have to be 0.3 times the mass of Earth to retain a subsurface ocean, or in other words only about three

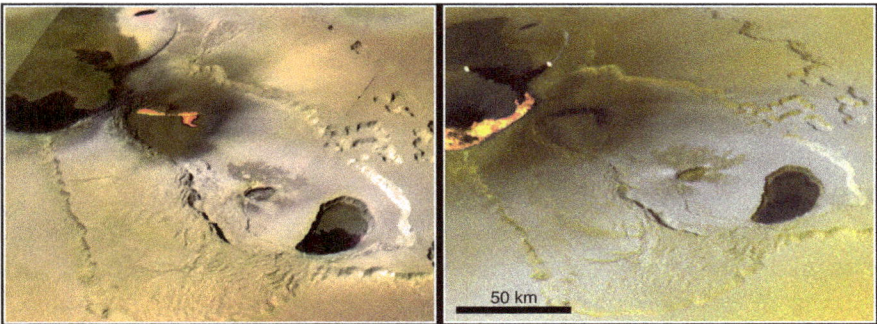

Fig. 1.2 Active volcanic flows on Io. The same region is shown here in the two images, but they were taken a few months apart and show substantial changes over this short period of time. Reprinted from https://www.jpl.nasa.gov. © 2000, NASA/JPL/University of Arizona. Public domain

times as big as Mars. In this case it would have a surface ice layer about 30 km thick.

A Steppenwolf planet with a large subsurface water ocean is one that could host life, a possibility that has been raised before regarding the subsurface ocean on Europa [79]. There are a number of possible life forms that could conceivably exist in such an environment, although our understanding of the possibly ecology of Europa or a Steppenwolf is limited, to say the least. Both ocean environments would be very dark, so life forms would have to obtain energy from sources other than light-based photosynthesis. This is not impossible as such organisms are plentiful on Earth, ranging from mushrooms to anaerobic bacteria to extremophiles, small organisms that live in extreme terrestrial environments that until recently were believed to be hostile to life. For example, extremophiles that live near hot water vents at the bottom of the ocean get their energy from heat and chemical reactions, a process known as chemosynthesis that does not require sunlight [80]. So in principle the dark oceans of Europa and Steppenwolf could be full of life. Some scientists have even gone so far as to speculate that oceans like these that are cocooned under a protective layer of ice will host life more frequently than the open seas of Earth-like planets, mostly because there are likely to be many more Europas and Steppenwolves than Earths [81]. There is yet another possible type of rogue planet that could maintain a subsurface ocean. A moon of a big rogue planet may experience tidal heating of the kind that helps warm Europa's ocean, so this is also a possible Steppenwolf environment. This is not guaranteed, though: since tidal heating requires a variation in the tidal forcing from one part of a Steppenwolf moon's orbit to another to rub the rocks against each other and warm them up, only moons with eccentric orbits would qualify. Tidal heating uses up energy and will gradually change the orbit of most Steppenwolf moons to a lower energy state, namely a circular orbit. Once the orbit has been circularized, there can be no more tidal heating unless the moon does not keep the same face towards its planet. Unless it had plenty of its own internal heat, the Steppenwolf moon would freeze solid.

We don't know exactly how many rogue planets are out there, but they certainly constitute an important class of planets. If one of them turns out to be relatively nearby, it will one day serve as a destination that will be a useful test of our ability to send spacecraft to the nearest stars. As others have already pointed out, the rogue planets could be our stepping stones to the stars of the known galaxy [82].

How Stars and Planets Are Named

A number of bright stars were given names in ancient times, like Sirius and Betelgeuse for example. Most of these names are of Arabic origin. More recently, a few well-known stars have been given modern names of their own. Other stars are given a variety of designations, as described below.

The Bayer System (1603) [83] In this designation, the brightest star in a chosen constellation is designated Alpha, the second brightest Beta, and so on through the 24 letters of the Greek alphabet, down to Omega. Sometimes the ordering by brightness is a little haphazard by modern standards, as in a number of constellations Alpha is not the brightest star. Sirius is the brightest star in the constellation Canis Majoris, so it is sometimes designated Alpha Canis Majoris. In most constellations, there are more than 24 stars visible to the naked eye, so Bayer continued his system by using Roman letters.

The Flamsteed System (1783) [84] The Flamsteed system uses numbers up to 100, although unlike the Bayer system, these are not arranged in order of brightness. In the Flamsteed system, Sirius is 9 Canis Majoris, although in the case of Sirius this label is hardly ever used.

There are also many star designations that consist of a catalogue name plus a number:

- The Histoire Céleste Francaise (1801) [85] is an "all-sky" catalogue that lists all stars down to stellar magnitude 9, or about 15 times dimmer than can be seen with the naked eye in a dark, rural sky. Its stars have the prefix "Lalande", after the famous French astronomer
- The Bonner Durchmusterung (1859) [86] (its stars are labelled BD plus a sky position identifier)
- The Henry Draper catalogue (1924) [87], whose stars are labelled with HD plus a number, with many nearby stellar systems usually referred to by their HD number and
- The Gliese nearby star catalogue (1957, with more recent updates) [88], with labels Gliese, GL or GJ plus a number

Some stars have been named after their discoverers: for instance, Barnard's Star, once a nickname but now the star's official name.

Planets are designated with a star name plus a lower case letter e.g. Barnard's Star b. For historical reasons, the planetary letters begin with "b" and there is no "a". Planets are designated in order of discovery, not distance from their star, so Barnard's Star b could be further from its star than Barnard's Star c. More recently, some well-known exoplanets have been given names of their own, and there is an international project that solicits suggestions for such names from the public (https://www.nameexoworlds.iau.org/).

How Extrasolar Planets Are Discovered

There are several astronomical techniques that have been used to find extrasolar planets, and each has their own advantages and disadvantages.

Direct Imaging This just means pointing a telescope at a star and seeing a planet in a photograph, either at visible or infrared wavelengths. Only a few large planets have been detected this way, as smaller extrasolar planets are very difficult or impossible to see with the telescopes currently available.

Transits This method requires measuring the light output of a star and seeing it decrease when a planet passes in front of the star (or "transits") and blocks its light. It is a simple method and even small amateur-sized telescopes have been used to find planets this way [89]. Careful measurements of transits can give a lot of information, like a good estimate of the size and orbit of the planet, plus even some clues to whether it has an atmosphere or not. A major drawback is that not every planet transits a star: many planets have orbits that are at a substantial angle to an imaginary line drawn between Earth and the star, and so do not transit. The angle of the orbit to this imaginary line is known as the "inclination". An orbit with an inclination of 0° is "face-on", with an observer from Earth appearing to look down on the orbit from above (see Fig. 1.3). An orbit that is "edge-on", with the orbit in the same plane as the imaginary line, has an inclination of 90°. The ideal situation to maximize the chance of a transit would be a planet whose orbit is exactly edge-on, as this would guarantee a transit. The worst case would be if the orbit were face-on, as this would mean no transits. Real planetary orbits are between these two extremes, with some planets transiting but most not doing so.

Radial Velocity The "radial" direction is the imaginary line between Earth and a star, as defined in the preceding paragraph. As a planet orbits a star, its gravity slightly pulls the star, first one way, and then the other [91]. The portion of the movement of the star that occurs in the radial direction is called the radial velocity. This back and forth motion causes slight changes in the

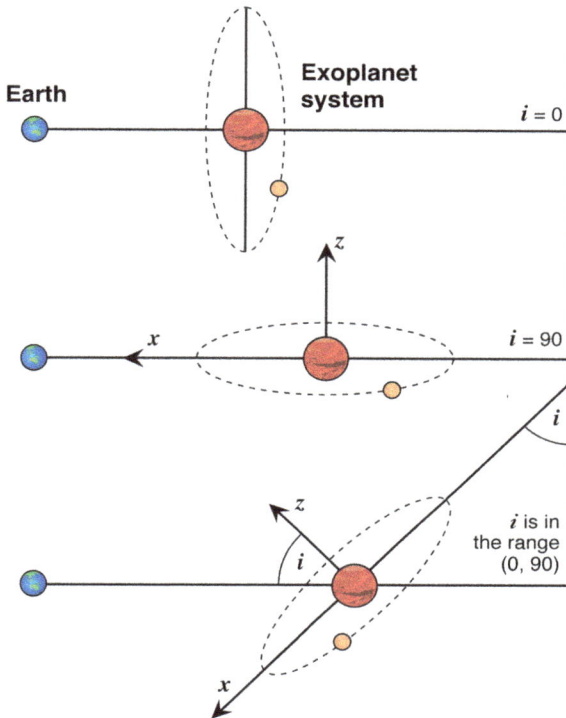

Fig. 1.3 The role of orbital inclination (denoted as the angle i) in whether an exoplanet transits across its star or not, relative to an imaginary line between the star and Earth. The top panel shows a "face-on" orbit, the center panel an "edge-on" orbit, while the bottom panel shows an orbit that is in between the two extremes. Modified with permission from [90]. © 1998, Cengage Learning. All rights reserved

frequency of the light emitted by it, in the same way that a Formula 1 car speeding past a grandstand causes those sitting there to hear a very noticeable decrease in the apparent pitch (i.e. frequency) of the vehicle's motor. This change of frequency is known as the Doppler effect. For stars, a simple way of saying this is that the tug of the orbiting planet can slightly change the color of the light emitted by the star, since red light has a lower frequency than blue light. The bigger the planet, the bigger the tug and the bigger the change in frequency. Making these measurements is not easy to do and only became routine in the 1990s. The problem with this method of discovery is that it doesn't tell us the exact mass of the planet. A particular frequency shift in the radial direction could either be caused by a small planet whose orbit is oriented edge-on to an observer on Earth, or the same shift could be caused by a very big planet whose orbit is almost face-on (see Fig. 1.3). To decide between the two, we need to know the inclination. We can get this information from transit observations, but not all planets transit. The radial velocity method only gives a minimum mass, which would be the case if the orbit were edge-on. The true mass would most often be larger than this.

The vast majority of extrasolar planets have been found using transits or radial velocity measurements, but there are some other techniques that have been used as well:

Astrometry [92] involves measuring the position of a star in the sky and observing the very small wobbles in that position caused by orbiting planets (see https://exoplanets.nasa.gov/resources/2288/astrometry/ for an animation that illustrates the method). This is a powerful technique in that it can give a lot of information about planetary orbits, but it works best for large worlds distant from a star. As the orbits of these worlds take a long time, so do the astrometry measurements needed to confirm the planet's existence. Also, the wobbles are very small and hard to measure.

Finally, *microlensing* [93] is a method that relies on planets passing in front of distant stars and the planet's gravity distorting the stellar image. The main problem with this technique is that it relies on luck, the random passage of a planetary system in front of a star. Follow-up observations are also usually impossible, as the typically microlensing planet might be thousands of parsecs away. Its main use is as a method to take a census of planetary types in the galaxy.

A note on exoplanet data. In this book, data on exoplanets such as size and distance from their star is taken from the Encyclopaedia of Planetary Systems (exoplanet.eu). Data for some systems is taken from NASA's Exoplanet Archive (https://exoplanetarchive.ipac.caltech.edu/) if there appears to be a significant difference between data sets. An attempt to reconcile the various data sets for the region within 10 parsecs has been made by Céline Reylé of CNRS in France and co-authors, and this will be referred to occasionally also [94].

References

1. The Martian (2015) imdb.com. https://www.imdb.com/title/tt3659388/?ref_=nv_sr_srsg_0_tt_8_nm_0_q_the%2520martian. Accessed 3 Jan 2023
2. Reylé C, Jardine K, Fouqué P, Caballero JA, Smart RL, Sozzetti A (2021) The 10 parsec sample in the Gaia era. Astronom Astrophys 650:A201

3. Quanz SP, Absil O, Benz W, Bonfils X, Berger JP, Defrère D, van Dishoeck E, Ehrenreich D, Fortney J, Glauser A, Grenfell JL (2021) Atmospheric characterization of terrestrial exoplanets in the mid-infrared: biosignatures, habitability, and diversity. Exp Astronom 54:1197–1221
4. Walker SI, Bains W, Cronin L, DasSarma S, Danielache S, Domagal-Goldman S, Kacar B, Kiang NY, Lenardic A, Reinhard CT, Moore W (2018) Exoplanet biosignatures: future directions. Astrobiology 18(6):779–824
5. Avatar (2009) imdb.com. https://www.imdb.com/title/tt0499549/?ref_=fn_al_tt_1. Accessed 5 Jan 2023
6. Baxter S (2012) The science of Avatar. Gollancz, London, p 118
7. NASA (2023) What makes a planet habitable? https://seec.gsfc.nasa.gov/what_makes_a_planet_habitable.html. Accessed 11 Jan 2023
8. Space Studies Board and National Research Council (2007) The limits of organic life in planetary systems. National Academies Press, Washington, DC
9. Hill ML, Bott K, Dalba PA, Fetherolf T, Kane SR, Kopparapu R, Li Z, Ostberg C (2023) A catalog of habitable zone exoplanets. Astronom J 165(2):34
10. Kopparapu RK, Ramirez R, Kasting JF, Eymet V, Robinson TD, Mahadevan S, Terrien RC, Domagal-Goldman S, Meadows V, Deshpande R (2013) Habitable zones around main-sequence stars: new estimates. Astrophys J 765(2):131
11. Ramirez RM (2018) A more comprehensive habitable zone for finding life on other planets. Geosciences 8(8):280
12. Kopparapu RK (2013) A revised estimate of the occurrence rate of terrestrial planets in the habitable zones around Kepler M-dwarfs. Astrophys J Lett 767(1):L8. https://doi.org/10.1088/2041-8205/767/1/L8
13. Schulze-Makuch D, Méndez A, Fairén AG, von Paris P, Turse C, Boyer G, Davila AF, Resendes de Sousa António M, Catling D, Irwin LN (2011) A two-tiered approach to assess the habitability of exoplanets. Astrobiology 11(10):1041–1052. https://doi.org/10.1089/ast.2010.0592
14. University of Puerto Rico at Arecibo (2024) Earth Similarity Index. https://phl.upr.edu/projects/earth-similarity-index-esi. Accessed 29 Jan 2024
15. Safonova M, Mathur A, Basak S, Bora K, Agrawal S (2021) Quantifying the classification of exoplanets: in search for the right habitability metric. Eur Phys J Special Topics 230:2207–2220
16. Rodríguez-Mozos JM, Moya A (2017) Statistical-likelihood Exo-Planetary Habitability Index (SEPHI). Mon Not Roy Astronom Soc 471(4):4628–4636
17. Prialnik D (2010) An introduction to the theory of stellar structure and evolution, 2nd edn. Cambridge University Press, Cambridge
18. Cannon AJ, Pickering EC (1912) Classification of 1,688 southern stars by means of their spectra. Ann Astronom Obs Harv Coll 56(5):115–164
19. van Leeuwen F (2007) Validation of the new Hipparcos reduction. Astron Astrophys 474(2):653–664. https://doi.org/10.1051/0004-6361:20078357

20. Repolust T, Puls J, Herrero A (2004) Stellar and wind parameters of Galactic O-stars. The influence of line-blocking/blanketing. Astron Astrophys 415(1):349–376. https://doi.org/10.1051/0004-6361:20034594

21. Kaler JB (2021) Stars: Zeta Ophiuchi. http://stars.astro.illinois.edu/sow/zetaoph.html. Accessed 3 Feb 2021

22. Ledrew G (2001) The real starry sky. J Roy Astronom Soc Can 95:32–33

23. Bond HE, Schaefer GH, Gilliland RL, Holberg JB, Mason BD, Lindenblad IW, Seitz-McLeese M, Arnett WD, Demarque P, Spada F, Young PA (2017) The Sirius system and its astrophysical puzzles: Hubble Space Telescope and ground-based astrometry. Astrophys J 840(2):70

24. Bonanno A, Schlattl H, Paternò L (2002) The age of the Sun and the relativistic corrections in the EOS. Astronom Astrophys 390(3):1115–1118

25. Cushing MC, Kirkpatrick JD, Gelino CR, Griffith RL, Skrutskie MF, Mainzer A, Marsh KA, Beichman CA, Burgasser AJ, Prato LA, Simcoe RA (2011) The discovery of Y dwarfs using data from the wide-field infrared survey explorer (WISE). Astrophys J 743(1):50

26. Luhman KL (2012) The formation and early evolution of low-mass stars and brown dwarfs. Ann Rev Astronom Astrophys 50:65–106

27. Vos JM, Burningham B, Faherty JK, Alejandro S, Gonzales E, Calamari E, Bardalez GD, Visscher C, Tan X, Morley CV, Marley M, Gemma ME, Whiteford N, Gaarn J, Park G (2023) Patchy forsterite clouds in the atmospheres of two highly variable exoplanet analogs. Astrophys J 944(2):138. https://doi.org/10.3847/1538-4357/acab58

28. Luhman KL (2013) Discovery of a binary brown dwarf at 2 pc from the Sun. Astrophys J Lett 767(1):L1. https://doi.org/10.1088/2041-8205/767/1/L1

29. Kirkpatrick JD, Gelino CR, Cushing MC, Mace GN, Griffith RL, Skrutskie MF, Marsh KA, Wright EL, Eisenhardt PR, McLean IS, Mainzer AK, Burgasser AJ, Tinney CG, Parker S, Salter G (2012) Further defining spectral Type "Y" and exploring the low-mass end of the field brown dwarf mass function. Astrophys J 753(2):156. https://doi.org/10.1088/0004-637X/753/2/156

30. Luhman KL (2014) Discovery of a ~250 K brown dwarf at 2 pc from the Sun. Astrophys J Lett 786(2):L18

31. Kopparapu RK, Hébrard E, Belikov R, Batalha NM, Mulders GD, Stark C, Teal D, Domagal-Goldman S, Mandell A (2018) Exoplanet classification and yield estimates for direct imaging missions. Astrophys J 856(2):122

32. Lammer H, Zerkle AL, Gebauer S, Tosi N, Noack L, Scherf M, Pilat-Lohinger E, Güdel M, Grenfell JL, Godolt M, Nikolaou A (2018) Origin and evolution of the atmospheres of early Venus, Earth and Mars. Astronom Astrophys Rev 26:1–72

33. Slade MA, Butler BJ, Muhleman DO (1992) Mercury radar imaging—evidence for polar ice. Science 258(5082):635–640. https://doi.org/10.1126/science.258.5082.635

34. Määttänen A, Montmessin F (2021) Clouds in the Martian atmosphere. In: Read P (ed) Oxford research encyclopedia of planetary science. Oxford University Press, Oxford. https://doi.org/10.1093/acrefore/9780190647926.013.114
35. Sorahana S, Yamamura I, Murakami H (2013) On the radii of brown dwarfs measured with AKARI near-infrared spectroscopy. Astrophys J 767(1):77. https://doi.org/10.1088/0004-637X/767/1/77
36. Wahl SM, Hubbard WB, Militzer B, Guillot T, Miguel Y, Movshovitz N, Kaspi Y, Helled R, Reese D, Galanti E, Levin S, Connerney JE, Bolton SJ (2017) Comparing Jupiter interior structure models to Juno gravity measurements and the role of a dilute core. Geophys Res Lett 44(10):4649–4659. https://doi.org/10.1002/2017GL073160
37. Fletcher LN, Helled R, Roussos E, Jones G, Charnoz S, André N, Andrews D, Bannister M, Bunce E, Cavalié T, Ferri F (2020) Ice giant systems: The scientific potential of orbital missions to Uranus and Neptune. Planet Space Sci 191:105030
38. Chen H, Rogers LA (2016) Evolutionary analysis of gaseous sub-Neptune-mass planets with MESA. Astrophys J 831(2):180
39. Zeng L, Jacobsen SB, Sasselov DD, Petaev MI, Vanderburg A, Lopez-Morales M, Perez-Mercader J, Mattsson TR, Li G, Heising MZ, Bonomo AS (2019) Growth model interpretation of planet size distribution. Proc Nat Acad Sci USA 116(20):9723–9728
40. Espinoza N, Brahm R, Jordán A et al (2016) Discovery and validation of a high-density sub-Neptune from the K2 mission. Astrophys J 830(1):43. https://doi.org/10.3847/0004-637X/830/1/4
41. Mocquet A, Grasset O, Sotin C (2014) Very high-density planets: a possible remnant of gas giants. Philos Trans Roy Soc A 372:20130164
42. Hébrard G, Lecavelier Des Étangs A, Vidal-Madjar A, Désert J-M, Ferlet R (2003) Evaporation rate of hot Jupiters and formation of chthonian planets. In: Beaulieu B, Lecavelier des Étangs A, Terquem C (eds) Extrasolar planets: today and tomorrow, ASP conference proceedings, Paris, France, 2003
43. Armstrong DJ, Lopez TA, Adibekyan V et al (2020) A remnant planetary core in the hot-Neptune desert. Nature 583(7814):39. https://doi.org/10.1038/s41586-020-2421-7
44. Alibert Y, Benz W (2017) Formation and composition of planets around very low mass stars. Astron Astrophys 598:L5
45. Fressin F, Torres G, Charbonneau D et al (2013) The false positive rate of Kepler and the occurrence of planets. Astrophys J 766:81
46. Tian Z, Magna T, Day JM et al (2021) Potassium isotope composition of Mars reveals a mechanism of planetary volatile retention. Proc Natl Acad Sci USA 118(39):e2101155118
47. Astronomy Magazine (2017) How big are the smallest spherical objects in our solar system? https://www.astronomy.com/science/how-big-are-the-smallest-spherical-objects-in-our-solar-system/. Accessed 18 Oct 2021

48. Griffith C, Mitchell JL, Lavvas P, Tobie G (2013) Titan's evolving climate. In: Mackwell SJ, Simon-Miller AA, Harder JW, Bullock M (eds) Comparative climatology of terrestrial planets. University of Arizona Press, Tucson, AZ, pp 1–27
49. Barclay T, Rowe JF, Lissauer JJ et al (2013) A sub-Mercury-sized exoplanet. Nature 494(7438):452–454. https://doi.org/10.1038/nature11914
50. Morbidelli A, Levison HF, Gomes R (2008) The dynamical structure of the Kuiper belt and its primordial origin. In: Barucci MA, Boehnhardt H, Cruikshank DP, Mordibelli A (eds) The solar system beyond Neptune. University of Arizona Press, Tuscon, AZ, pp 275–292
51. Brown ME (2012) The compositions of Kuiper belt objects. Annu Rev Earth Plan Sci 40:467–494
52. NASA (2018) 10 things to know about the Kuiper belt. https://science.nasa.gov/solar-system/kuiper-belt/10-things-to-know-about-the-kuiper-belt/. Accessed 31 Jan 2024
53. Brown ME, Trujillo C, Rabinowitz D (2004) Discovery of a candidate inner Oort cloud planetoid. Astrophys J 617(1):645
54. Trujillo CA, Sheppard SS (2014) A Sedna-like body with a perihelion of 80 astronomical units. Nature 507(7493):471–474
55. Brown ME, Batygin K (2021) The orbit of Planet Nine. Astronom J 162(5):219. https://doi.org/10.3847/1538-3881/ac2056
56. Stevenson DJ (1999) Life-sustaining planets in interstellar space? Nature 400:32
57. Alfè D, Gillan MJ, Price GD (2007) Temperature and composition of the Earth's core. Contemp Phys 48(2):63–80. https://doi.org/10.1080/00107510701529653
58. Davies JH, Davies DR (2010) Earth's surface heat flux. Solid Earth 1(1):5–24. https://doi.org/10.5194/se-1-5-2010
59. Covey C, Haberle RM, McKay CP et al (2013) The greenhouse effect and climate feedbacks. In: Mackwell SJ, Simon-Miller AA, Harder JW, Bullock M (eds) Comparative climatology of terrestrial planets. University of Arizona Press, Tucson, AZ, pp 163–179
60. Goldblatt C, Watson AJ (2012) The runaway greenhouse: implications for future climate change, geoengineering and planetary atmospheres. Philos Trans Roy Soc A 370(1974):4197–4216
61. Pierrehumbert R (2010) Principles of planetary climate. Cambridge University Press, Cambridge, p 512
62. Way MJ, Del Genio AD, Kiang NY et al (2016) Was Venus the first habitable world of our solar system? Geophys Res Lett 43(16):8376–8383
63. Kopparapu RK, Ramirez R, Kasting JF et al (2013) Habitable zones around main-sequence stars: new estimates. Astrophys J 765(2):131
64. Parks J (2019) Planet Nine may be a black hole the size of a baseball. https://astronomy.com/news/2019/10/planet-nine-may-be-a-black-hole-the-size-of-a-baseball. Accessed 3 Apr 2021

65. Brown K, Mathur H (2023) Modified Newtonian dynamics as an alternative to the Planet Nine hypothesis. Astronom J 166(4):168
66. Nguyen MM, De Rosa RJ, Kalas P (2020) First detection of orbital motion for HD 106906 b: A wide-separation exoplanet on a Planet Nine-like orbit. Astronom J 161(1):22
67. Alibert Y, Carron F, Fortier A et al (2013) Theoretical models of planetary system formation: mass vs. semi-major axis. Astronom Astrophys 558:A109
68. Strigari LE, Barnabè M, Marshall PJ et al (2012) Nomads of the galaxy. Mon Not Roy Astronom Soc 423(2):1856–1865. https://doi.org/10.1111/j.1365-2966.2012.21009.x
69. Lingam M, Hein AM, Eubanks TM (2023) Chasing nomadic worlds: A new class of deep space missions. Acta Astronaut 212:517–533
70. Sumi T, Koshimoto N, Bennett DP et al (2023) Free-floating planet mass function from MOA-II 9 yr survey toward the galactic bulge. Astronom J 166(3):108
71. Nesvorný D (2011) Young solar system's fifth giant planet? Astrophys J Lett 742(2):L22. https://doi.org/10.1088/2041-8205/742/2/L22
72. Abbot DS, Switzer ER (2011) The Steppenwolf: a proposal for a habitable planet in interstellar space. Astrophys J Lett 735(2):L27
73. Hesse H (1927) Der Steppenwolf. S Fischer, Berlin
74. Thomas PC, Tajeddine R et al (2016) Enceladus's measured physical libration requires a global subsurface ocean. Icarus 264:37–47. https://doi.org/10.1016/j.icarus.2015.08.03
75. Park RS, Bills B, Buffington BB (2015) Improved detection of tides at Europa with radiometric and optical tracking during flybys. Plan Space Sci 112:10–14. https://doi.org/10.1016/j.pss.2015.04.005
76. Mareschal JC, Jaupart C (2021) Energy budget of the Earth. In: Gupta HK (ed) Encyclopedia of solid earth geophysics. Springer, Cham, pp 361–368
77. Keane JT, Matsuyama I, Bierson CJ, Trinh A (2023) Tidal heating and the interior structure of Io. In: Lopes RMC, de Kleer K, Keane JT (eds) Io: A new view of Jupiter's moon. Springer, Cham, pp 95–146
78. Vilella K, Choblet G, Tsao WE, Deschamps F (2020) Tidally heated convection and the occurrence of melting in icy satellites: application to Europa. J Geophys Res Planets 125(3):e2019JE006248
79. Chyba CF, Phillips CB (2002) Europa as an abode of life. Orig Life Evol Biosphere 32(1):47–67
80. Dick GJ (2019) The microbiomes of deep-sea hydrothermal vents: distributed globally, shaped locally. Nat Rev Microbiol 17(5):271–283
81. Southwest Research Institute (2021) Worlds with underground oceans may be more conducive to life than worlds with surface oceans like Earth. ScienceDaily, 16 Mar 2021. https://www.sciencedaily.com/releases/2021/03/210316100719.htm. Accessed 20 Mar 2021

82. Glister P (2022) Chasing nomadic worlds: Opening up the space between the stars. https://www.centauri-dreams.org/2022/12/28/chasing-nomadic-worlds-opening-up-the-space-between-the-stars/. Accessed 31 Jan 2024

83. Wikipedia (2023) Bayer designation. Accessed 10 Dec 2023

84. Wikipedia (2023) Flamsteed designation. Accessed 10 Dec 2023

85. Wikipedia (2023) Histoire Céleste Francaise. Accessed 10 Dec 2023

86. Wikipedia (2023) Bonner Durchmusterung. Accessed 10 Dec 2023

87. Wikipedia (2023) Henry Draper Catalogue. Accessed 10 Dec 2023

88. Wikipedia (2023) Gliese catalogue of nearby stars. Accessed 10 Dec 2023

89. Naeye R (2004) Amateur detects exoplanet transit. In: Sky and Telescope. https://skyandtelescope.org/astronomy-news/amateur-detects-exoplanet-transit/. Accessed 5 Feb 2021

90. Zeilik M, Gregory SA (1998) Introductory astronomy and astrophysics. Cengage Learning, Boston, MA, p 236

91. Mayor M, Queloz D (1995) A Jupiter-mass companion to a solar-type star. Nature 378(6555):355–359

92. Wikipedia (2023) Astrometry. Accessed 10 Dec 2023

93. Wikipedia (2023) Gravitational microlensing. Accessed 10 Dec 2023

94. Reylé C, Jardine K, Fouqué P et al (2021) The 10 parsec sample in the Gaia era. Astronom Astrophys 650:A201

2

Chiron's Place

The Alpha Centauri Sector: 12–18 h, 0 to –90°

Our tour of the known galaxy continues in a part of the night sky that is visible largely from the Southern Hemisphere (see Fig. 2.1). For the purposes of the tour, I've divided the night sky into eight sectors, delineated by an east-west coordinate system of 24 "hours", known to astronomers as right ascension, and a north-south coordinate system divided into 180°, known as declination. Here, the Alpha Centauri sector runs from 12 to 18 h right ascension and zero to –90° declination. Figure 2.1 is plotted a little like a Mercator projection of the Earth's surface, a rectangular mapping that makes the polar regions look enormous and that mistakenly convinced a whole generation of schoolchildren that Greenland was almost as big as Africa. The polar regions of Fig. 2.1 are distended as a result: for instance, the southern polar constellation Octans is really a lot smaller than shown.

The best-known constellations in this sector are no doubt the zodiacal ones of Scorpius, Libra and Virgo. Scorpius is dominated by the red supergiant Antares, whose name means "a rival to Mars", a reference to the brightness and redness of both objects. Antares is about 170 parsecs away, so for our purposes is not nearby and not part of the known galaxy [1]. The constellation Libra does not have any very bright stars, although its brightest, the Alpha Librae system, is only about 23 parsecs away and consists of at least four stars, two pairs each orbiting each other. The constellation Virgo includes the bright star Spica about 77 parsecs away, two large blue stars zooming around each other so closely that their mutual gravity has distorted them into egg shapes rather than spheres. This sector also boasts a part of Ophiuchus, the constellation Crux (otherwise known as the Southern Cross), and a number of dim

© The Author(s), under exclusive license to Springer Nature Switzerland AG 2024
K. J. E. Walsh, *Planets of the Known Galaxy*, Science and Fiction,
https://doi.org/10.1007/978-3-031-68218-6_2

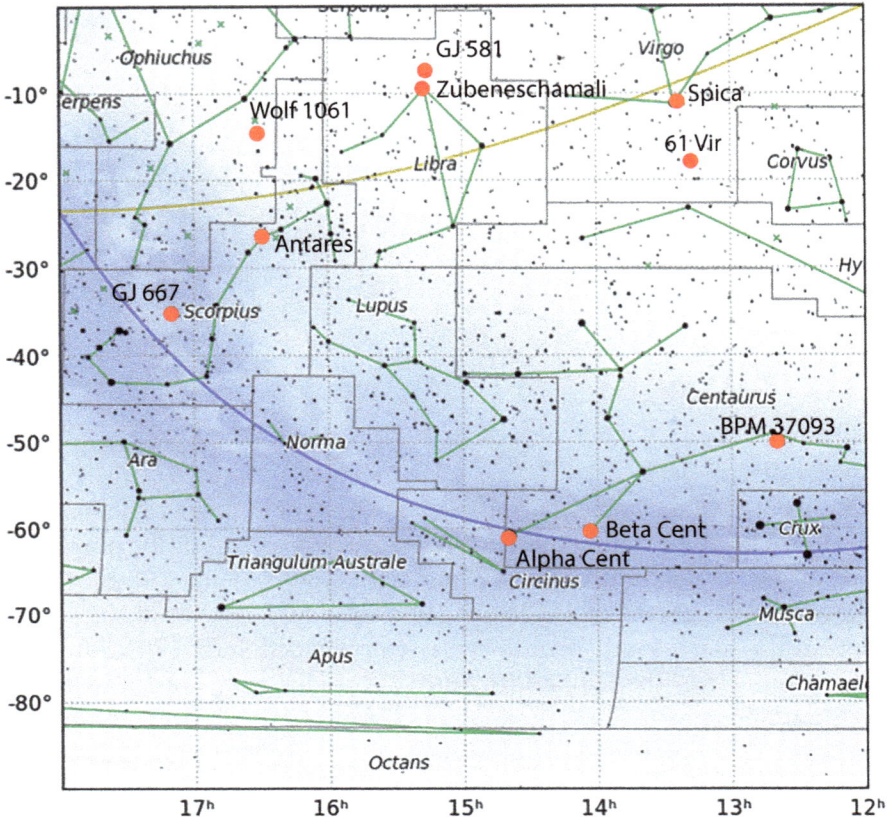

Fig. 2.1 Star map of the Alpha Centauri sector. Blue shading indicates the "Milky Way", the most star-packed part of the galaxy, while the blue (darker) line is the galactic equator, an imaginary circle running through the Milky Way that slices the galaxy in half. The orange (lighter) line is the path of the Sun's location through the seasons (indicating the "zodiac" constellations). Larger circles indicate brighter stars, while locations mentioned in this chapter are indicated by red circles. Image produced using StarCharter (https://github.com/dcf21/star-charter, accessed on the 2nd of July 2024. © 2020, Dominic Ford) software under the GNU general public license

minor constellations in the far southern sky that are invisible from the Northern Hemisphere (like Octans).

This sector also contains the large constellation Centaurus. As its name suggests, it represents the mythical centaurs, creatures that are half man, half horse, renowned in Greek and Roman mythology as highly rowdy drunks. But not all centaurs were party animals: Ovid, the Roman poet, claimed that the constellation Centaurus actually depicted the centaur Chiron, a studious type who was a tutor of Hercules and many other legendary figures of Greek culture, and who was a well-known expert in ancient medicine [2, 3]. The brightest star in Centaurus and in this whole sector is Alpha Centauri, and despite its location in the southern sky, the ancients of the Northern Hemisphere knew it well. The Egyptian-based Greek astronomer Ptolemy lists it in his second-century star catalogue [4], and in those days it could be briefly seen as far north as Greece, a brilliant harbinger of the unknown skies below the southern horizon. But since then, the slow wobbles in the orientation of Earth's axis that help cause the ice ages have pushed the apparent position of Alpha Centauri in the sky further south. At present, only those living south of about Galveston, Texas can see it. In the Southern Hemisphere it is a potent symbol, as together with Beta Centauri it points towards the Southern Cross, an emblem on the flags of several Southern Hemisphere nations. Alpha Centauri has always been a little special, but it was only long after it was first identified that astronomers discovered exactly how special it was.

The Alpha Centauri system is the closest known extrasolar system to Earth. Distances to nearby stars are determined by geometric measurements. As the Earth revolves around the Sun, we see stars from one side of the Sun and then 6 months later from the other side. Thus nearby stars appear to shift slightly in position in the sky over the course of a year. The shift, known as the "parallax", is tiny but can be measured for many of the closer stars. By the early part of the nineteenth century, the instruments and telescopes had become available to allow these measurements to be made. In 1839, astronomer Thomas Henderson of Cape Town announced to the world that Alpha Centauri had the largest parallax ever measured [5]. These observations had been made years before, but Henderson did not believe them because the shift was so big that it meant that Alpha Centauri was the closest star. One day we might find a very dim star that is closer, but for now the Alpha Centauri system is still the nearest.

I am careful to say "the Alpha Centauri system", because this system consists of more than one star. It was astronomer Father Jean Richaud, a Jesuit priest, who in 1689 used his trusty 12-ft telescope set up in Pondicherry, India, to discover that Alpha Centauri was actually two stars instead of one [6]. Amateur astronomy at remote locations was a difficult task in those days, because during his travels through Asia, several of Father Jean's colleagues died [7]. The two stars of Alpha Centauri received their current names in 2018 from the international committee that decides such things and are now called Rigel Kentaurus and Toliman. Rigel Kentaurus is still also known as Alpha Centauri A, while Toliman is known as Alpha Centauri B. They circle each other in an oval-shaped orbit, with the separation between them varying from about the distance from the Sun to Neptune to about the distance from the

Sun to Saturn. In 1915, a much fainter third star in this system was discovered by Robert Innes at Johannesburg Observatory, known as Proxima Centauri, meaning "nearest Centaur" because of its proximity to our solar system [8]. By the standards of our own solar system, though, Proxima is a very long way from Alpha Centauri A and B: about 1.8 trillion kilometers or about 400 times the distance from the Sun to Neptune, or about 13,000 a.u. Proxima's parallax indicates that it is slightly closer to our solar system than the rest of the Alpha Centauri system, so Proxima is the closest known star to Earth. This fact alone would be enough to make Proxima an object of intense interest. But in 2016, the European Southern Observatory, situated on a desert mountain in northern Chile, announced that Proxima has a planet [9].

Known as Proxima Centauri b, the planet was discovered by the radial velocity method (see Chap. 1). No transits have been observed, so at present we only know the minimum mass of the planet, recently measured at 1.17 Earth masses [10]. But we can say more about the actual mass of Proxima Centauri b from studies of other stars like it. Proxima is an M star, considerably smaller and dimmer than a G star like the Sun. Some planets of M stars have masses that are known much more precisely than Proxima's because these planets have both radial velocity and transit measurements. It has been found that for M stars like Proxima, the most likely planetary mass is in the range of about one to seven times the mass of Earth, as M stars do not usually have very large planets like Jupiter with 300 times the mass of Earth. Comparing the spread of these previous planetary mass estimates for M stars to the observed minimum mass measurements for Proxima, Alex Bixel and Daniel Apai of the University of Arizona concluded in 2017 that the most likely actual mass of planet b is about 1.6 Earth masses, with a radius of about 1.1 Earth radii [11]. Later work has suggested a mass of about 2.1 Earth masses and that therefore the planet is most likely a smallish super-Earth [11, 12]. We don't know this for sure; there is still a chance that it might be a larger sub-Neptune world. But let's go with the most likely outcome and see if there is more that can be said about Proxima Centauri b. Before we do that, though, instead of calling this planet "Proxima Centauri b" all the time, which is a bit of a mouthful, let's call it Chiron's World.

A starting point for a more informed description of Chiron's World is to estimate its temperature. Since we know the distance of the planet from its star and we know the brightness of the star, it is easy to calculate the total amount of stellar radiation that it receives. This is about 60% of the amount that Earth receives from the Sun, making Chiron's World a "warm super-Earth" based on the classification introduced in Chap. 1. Its equilibrium temperature is between about 230 K (−43 °C; −45 °F) and about 251 K (−22 °C;

−8 °F), similar to Earth's equilibrium temperature of about 255 K (−18 °C; 0 °F) [13]. This relatively mild temperature for Chiron's World is promising, but to determine the evolution of its climate we also need to know how hot it was in the distant past. M stars go through a phase early in their evolution when they are much brighter than they are in later life. This would not be good news for a planet like Chiron's World, as during this early phase of Proxima's development, Chiron's World could have become so hot that almost all of its water could have been boiled away and lost to space [14]. The planet might have ended up as a bone-dry furnace, like Venus today. Even so, let's assume for the sake of argument that Chiron's World is not just a bigger, uglier version of Venus, and instead has a relatively thin atmosphere like Earth's and a modest greenhouse effect.

We also know that computer simulations of planetary formation around small stars like Proxima mostly produce planets that are completely water-covered [15]. The reason is related to the way planets form. Planets coalesce from bits of rock and ice that orbit a star. Sometimes a lot of ice is collected, sometimes very little. Thus planets can end up with very different water contents: they might end up with less water than Mars, or they could collect so much water that there is absolutely no reason why they have to end up with any land surface area at all. This is particularly the case for planets that form in cold, icy regions of a stellar system, collect a lot of ice, and then migrate inwards. Recent simulations strongly suggest that Chiron's World formed like this, and so it is highly likely to have a large water content [16]. We will call that proposed planet Chiron's Waterworld.

Another major difference from terrestrial conditions is the length of this planet's day. Chiron's Waterworld is likely to keep the same face towards Proxima all the time, due to the same phenomenon that keeps the same part of the Moon always facing Earth, known as "synchronous rotation". Anyone who occasionally has had a good look at Earth's Moon will notice that it appears to have similar features every time, the suggestion of a distorted face, the so-called Man in the Moon. Since the Moon looks upside down in the Southern Hemisphere compared to its appearance in the Northern Hemisphere, the familiar Northern Hemisphere scowl of the Man in the Moon has the frown turned upside down in regions south of the equator, making him look much more cheerful. Irrespective, the reason that the Man in the Moon is always visible when the Moon is close to full is that the Moon always has more or less the same side facing the Earth, as the time taken for the Moon to rotate once on its axis is exactly the same as the time required for the Moon to make one complete orbit around the Earth. Thus the Moon has a solar "day" (or "sol") of about 28 Earth days.

This situation is no accident, as it is due to the tidal heating caused by the Earth pulling on the Moon. The energy required to create this heat has to come from somewhere, and it comes from energy loss due to the gradual slowdown of the Moon's rotation speed. Eventually, billions of years ago, the rotation rate settled down into a situation where it is harder to extract further energy. This low-energy situation occurs because the Moon is slightly American-football-shaped, and so the pointy, more massive bit of the football now faces Earth all the time, apart from some minor side-to-side wobbling known as libration caused by the Moon's slightly eccentric orbit. If the massive bit of the Moon did not face the Earth, the Earth's gravity would pull on the Moon until it did, a bit like a magnet pulling on a compass needle, or like a heavy, oblong object on a hill eventually ending up with the heaviest bit pointing downhill. Since the Moon used to be closer to the Earth and because the Earth is much larger than the Moon, this process of "tidal locking", as it is called, only took a few million years [17]. Now of course the Moon is also pulling on the Earth, but so far it has only succeeded in slowing Earth's rotation rate somewhat, but not slowing it to the same as the Moon's. Originally, the Earth's rotation speed was set by a glancing collision about 4.5 billion years ago between the Earth and a Mars-sized ancient planet that no longer exists, an impact that formed the Moon. Calculations suggest that just after this event the Earth would have rotated about once every 5 h (or even less) instead of today's 24 h [18, 19]. The slowdown since then is due to the Moon's gravity.

This slow decrease in the Earth's rotation rate can be measured accurately with modern techniques, but strangely enough it also can be estimated using ancient records of eclipses. These show that ancient solar eclipses would have occurred in the wrong locations if we assume that the rotation rate of the Earth is the same as it is today [20]. Instead, it has slowed by about 1.8 thousandths of a second per century in the past 3000 years. A bigger slowdown can also be seen in ancient fossils. Some fast-growing shells from the Mesozoic Era about 70 million years ago have growth rings that were formed every day, and since there are also seasonal patterns to shell growth, the number of days per year can be calculated. This gives a Mesozoic day length of 23.5 h instead of today's approximately 24 h [21]. In any event, the importance of this process for extrasolar planets is that planets in close orbits around stars can experience an even greater tidal locking effect than that of the Earth on the Moon. Those planets are very likely to always have the same face towards their star. This can have big implications for their climates, as the side that always faces the star will be strongly heated all the time. Another implication is that there is nothing special about a day length of 24 h for an Earth-like planet. Since collisions between planets and other bodies that are large enough to change a planet's rotation rate can happen during the formation of a solar system, the day length on a habitable Earth-like planet could be practically anything, from a few hours to hundreds of Earth days. In the case of Chiron's Waterworld, synchronous rotation implies that its rotation would take the same amount of time as it takes to orbit Proxima, about 11.2 days.

As a result, the climate of Chiron's Waterworld would be quite different from Earth's. Climate is the average weather of a planet over a long period of time, and it can be simulated by climate models, large computer codes that have previously been used to simulate the climate of the Earth. These models work by simultaneously solving various equations that represent the physical processes that take place in the ocean and atmosphere. The result is the spontaneous generation of weather systems and ocean currents that represent an approximation of the actual observed atmosphere and ocean. The weather itself is not coded in; instead, it is calculated from the equations, in a very similar way to how weather forecast models generate tomorrow's weather. This makes climate models suitable for simulating the climate of planets other than the Earth, as the required input variables such as the amount of incoming solar radiation and the planetary rotation rate can be changed in the model and a new simulation can then be performed. This results in different weather and a different climate.

All of this sounds straightforward in principle, but the reality is that painstaking work has been undertaken over several decades by various research groups worldwide to gradually improve the simulations of these models. In addition, the models used for Earth's climate require some modifications before they can be used for simulating the climates of other planets, especially to their calculations of atmospheric infrared radiation [22]. Even so, these modified models are increasingly being used to estimate the potential range of climates of extrasolar planets, as a way of determining how habitable they might be, as well as to simulate the possible climates of individual planets. Not only do the climate models assist us to build our knowledge of likely conditions on individual planets, the models also help astronomers to focus on those planets that are most likely to be habitable.

A climate simulation of a water-covered, synchronously rotating Chiron's Waterworld has already been performed by a team led by climate scientist Anthony Del Genio of NASA [23]. Since the amount of stellar radiation received by Chiron's Waterworld is considerably less than that received by Earth, the resulting simulated climate is cold: assuming a greenhouse effect similar in size to Earth's, the maximum simulated average surface temperature is only 3 °C (37 °F), occurring in the region of the planet where the star is directly overhead all the time (known as the "sub-stellar point"). This temperature is similar to that in parts of our Arctic Ocean during summer. The slow rotation of Chiron's Waterworld causes some unusual effects on its pattern of winds and currents, and thereby the geographic pattern of its surface temperatures. The result is a so-called "lobster" pattern (Fig. 2.2 shows a similar pattern for a different planet). Oddly enough, a similar lobster pattern is induced in the equatorial Pacific during El Niño conditions: meteorologists know this as the Matsuno-Gill response, after the two atmospheric scientists who in the 1960s came up with the theory that predicted it [24]. On Chiron's Waterworld, due to warm ocean currents that efficiently transport heat away from the day side, the planet has some open water even on the night side. Nevertheless, there are very cold temperatures on the night side near the poles, typically around −60 °C, and the sea there is completely ice-covered. Simulations where the size of the planet's greenhouse effect was substantially increased only ended up increasing the substellar point temperature by a few degrees. The reason for this is that Proxima, being an M star, emits radiation mostly at infrared wavelengths. This radiation is more easily absorbed by the greenhouse gases in the atmosphere than visible radiation, which largely passes through the atmosphere and so tends to warm the solid surface instead. This means that a lot of the radiant energy from Proxima is going into heating the higher layers of the atmosphere rather than the surface. Importantly, this

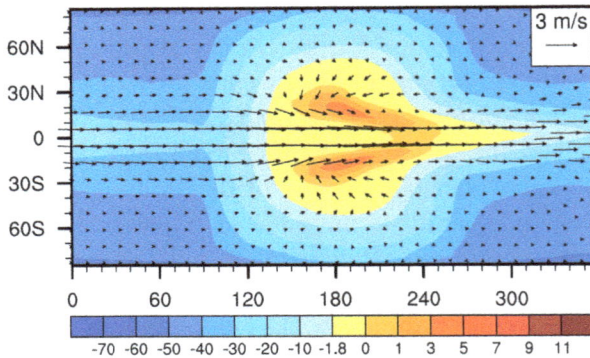

Fig. 2.2 Surface air temperature (°C) from a waterworld simulation of the climate of GJ 581 g. The arrows indicate surface current velocity. The horizontal axis is longitude in degrees, and the vertical axis is latitude. The substellar point is at the equator and 180° longitude. Reprinted with permission from [25]. © 2014, National Academy of Sciences. All rights reserved

weakens the impact of the greenhouse effect on surface temperatures for Chiron's Waterworld and other M star planets.

This implies that the main limitation on day side temperatures on these worlds is the meagre amount of radiation provided by Proxima or other M stars. As only some of this radiation is visible light, the light levels in the equatorial regions of Chiron's Waterworld would be low, only about a fiftieth of Earth's. Climate simulations also suggest that the sub-stellar regions on Chiron's Waterworld would be quite cloudy, which combined with the niggardly polar-like illumination would make the place rather gloomy [23]. This type of climate, a kind of perpetual Arctic summer in a perpetual twilight, would likely be common among exoplanets of M stars for this level of incoming radiation, even for very different greenhouse gas concentrations, provided that the atmosphere is relatively thin like Earth's. Much bigger atmospheres with much larger greenhouse gas concentrations would be required to heat up the surface substantially.

Chiron's Waterworld is an Arctic world, with about 60% of its ocean covered by sea ice. If some land were included, simulations show that the habitability of the planet would be improved slightly: mean temperatures on land regions at the sub-stellar point could then be as high as 30 °C (86 °F), and the surrounding ocean would warm up a little as well [26]. Still, most of the planet would remain ice-covered. We also know from these simulations that Chiron's Waterworld would be very windy, with typical average wind speeds on the day side near the substellar point of 10–15 ms^{-1} (20–30 knots). This is strong enough to make the wind difficult to walk against most of the time and is considerably windier than the vast majority of places on Earth's land surface. These strong winds are largely caused by the substantial temperature contrast between the perpetual day and perpetual night of the two halves of the planet, as the hotter air from the day side rises and colder air from the night side continually rushes in to replace it.

The climate simulations outlined above assume that Chiron's Waterworld has one side with perpetual day and the other side with perpetual night. While this synchronous rotation case is the most likely orbital configuration for Chiron's Waterworld, it is not the only possible one. For instance, in our solar system, Mercury is close enough to our Sun to be forced into synchronous rotation, but it hasn't been. Instead, it rotates three times for every two times it orbits the Sun, known technically as a "3:2 spin-orbit resonance". There are two reasons for this. The first is that like all planets, the mass inside Mercury is not entirely evenly distributed: instead, there is a lumpy, more massive bit that will tend to face "downhill" towards the Sun. Mercury's eccentric orbit means that the Sun's tidal force will vary from one part of its orbit to another depending on the variation of its distance from the Sun. This enables the lumpiest bit of Mercury's interior to point in different directions rather than always towards the Sun, thus liberating it from synchronous rotation. The implication is that if the orbit of Chiron's Waterworld were eccentric, it might not be rotating synchronously. Its eccentricity is currently not known precisely, but it is probably less than Mercury's [27]. Nevertheless, if it is large enough to create a 3:2 spin-orbit resonance, this would create a different planetary climate from the synchronous rotation case. Its sol would then be about 22 Earth days and the sub-stellar point would appear to move around the planet's equatorial regions rather than just stay in the same location all the time. This would lead to a large equatorial band of open water around the entire planet [23].

The idea of Chiron's Waterworld as an Arctic ocean world does not sound too bad, but there is another serious potential limitation on its habitability. Proxima has occasional flares, or explosions on its surface, that can considerably increase its brightness. Flares are caused by variations in the star's magnetic field, fluctuations that release energy and then push electrons and other particles outwards. The particles then collide with the star's atmosphere, releasing heat in the form of a flare. Like planets, stars that spin faster have stronger magnetic fields. Young M stars spin faster and have much stronger magnetic fields than older G stars like the Sun, so they experience more intense flares [28]. Another reason that M stars have more flares than G stars might be related to differences in the way that energy is transported within the stars. In the core of a G star, energy is transported by radiation, but closer to the surface, it is transported by convection, the turbulent movement of hot gas rocketing upwards, a little like a hot air balloon. In contrast, in some of the smaller M stars, energy transport throughout the entire star is performed by convection alone. It is thought that this difference would have an effect on the characteristics of a star's magnetic field, although this appears to be an active area of research [29].

On Proxima, small flares typically occur about twice a day; less frequently, about several times a year, there are much bigger ones. In 2018, a particularly large flare caused Proxima to become about 70 times brighter at visible wavelengths than it normally is, causing the star for a few minutes to appear brighter at the surface of Chiron's Waterworld than our Sun does on Earth [30]. In 2021, a truly enormous flare increased the ultraviolet (UV) light from Proxima by about 10,000 times [31]. Flares are rich in UV, and the estimated UV flux at the surface of Chiron's Waterworld during this flare was a lot higher than the typical UV flux at the surface of Earth. Earth's surface is partly protected from UV by its ozone layer, a region of the upper atmosphere where complicated reactions turn oxygen with two atoms (O_2, hopeless at absorbing UV) into ozone (O_3, which can absorb UV and so provides some protection against it). Studies have suggested that the cumulative effect of Proxima's flares would be to remove any ozone layer that Chiron's Waterworld might possess within less than a million years, an eyeblink in time compared to its planetary age of several billion years [30]. This would be a serious issue for any land-based life there but would probably not really be a big problem for sea life, as UV is very effectively absorbed by sea water and does not penetrate much more than a few tens of meters below the surface [32]. Any large flares might cause some dieback of photosynthesizing organisms living close to the surface of the ocean, but between big flares these regions could likely be recolonized by organisms from deeper waters.

All of this assumes that Chiron's Waterworld actually has any atmosphere at all. Large flares are associated with bursts of stellar material flying outward and colliding with the atmosphere of the planet. If these collisions are large enough and frequent enough, they could entirely strip away its atmosphere. On the other hand, as we mentioned in Chap. 1, if the planet has a strong magnetic field to protect it, atmospheric loss would be considerably less. For Chiron's Waterworld, the SEPHI index calculates that having such a strong magnetic field is less likely because its strength depends upon the planet's rotation rate, with a faster rotation rate giving a stronger magnetic field. Assuming synchronous rotation [33], a sol of 11 days implies a weaker magnetic field. This is a potential problem, as recent calculations have shown that for an Earth-like planet with a magnetic field as strong as Earth's at the distance of Chiron's Waterworld from Proxima, the atmospheric loss due to stripping would be enough to remove an atmosphere as thick as Earth's in only about 300 million years [34]. This is not a deal breaker, though, as the replenishment of Proxima's atmosphere by volcanic activity could easily be fast enough to counteract the stripping effect. Right now, there are too many variables here to say for sure, but it is clear that atmospheric stripping will make the atmosphere of Chiron's Waterworld thinner than it would otherwise be.

Even if Chiron's Waterworld has no land, and even if it is cold and gloomy, it would still be more habitable than any other planet in our solar system except Earth. Why? We are first assuming here that the atmosphere of Chiron's Waterworld is thick enough that space suits will not be needed, and instead that scuba-type gear would be sufficient. Temperatures would be cool but this could be easily dealt with by protective clothing. On no planet in our solar system apart from Earth could this level of protection be sufficient to preserve the life of an astronaut.

The allure of our nearest stellar system has meant that modern authors of speculative fiction have been writing about Alpha Centauri for decades. In Larry Niven's Known Space series, *The Man-Kzin Wars* (1988 and following volumes, many written by other authors) [35], humans fight a guerilla campaign against their feline enemies the Kzin on the idyllic planet Wunderland, circling Alpha Centauri A. Descriptions of the climate of Wunderland indicate that it is a benign, warm world, with the exception of a southern continent that has a cool, dry climate a little like that of Patagonia. When the K star Alpha Centauri B is above the horizon, the sky turns purple. More recently, the film *Avatar* (2009) takes us to Pandora, an Earth-like moon of the giant planet Polyphemus in orbit around Rigil Kentaurus [36]. Pandora has a largely tropical climate, perhaps because its atmosphere has a high concentration of the greenhouse gas carbon dioxide, a concentration so high that it is unbreathable by human beings [37]. Pandora is smaller than Earth and so has a weaker surface gravity. Usually, due to the giant planet's enormous tides, we would assume that Pandora would always keep the same face towards Polyphemus. Thus the rising and setting of Alpha Centauri A should be dictated by the time taken for Pandora to orbit Polyphemus, which based on analogy to the moons of Jupiter must be at least a couple of Earth days. On the other hand, in the movie, the planet has a sol of 26 standard hours [38]. Also, due to tidal heating, Pandora is a lot more volcanic than Earth. No planets have been yet confirmed around Alpha Centauri A. But in early 2021, in a direct image taken from a large telescope, a blip was observed that is consistent with a sub-Jovian or Jovian planet orbiting about 1.1 a.u. away from this star [39]. Polyphemus, anyone?

There may be other planets in the Alpha Centauri system yet to be discovered. In 2020, a group of astronomers at the University of Texas performed a sophisticated analysis of the available radial velocity data to conclude tentatively that there was a second planet orbiting Proxima, Proxima Centauri c [27]. This world, if it really exists, is considerably further out and considerably larger than Chiron's Waterworld and is likely a very cold sub-Neptune. It should be visible in large telescopes using special techniques, and indeed initial attempts to see it have found something that resembles a planet plus what might be a gigantic ring system larger than Saturn's [40]. Even more recently, in 2022 astronomers announced the discovery of Proxima Centauri d, a hot rocky world orbiting Proxima every 5 days, about which very little is currently known [41].

Older tales about the constellation Centaurus put Alpha Centauri next to or on top of one of Chiron's feet. This is ironic, as according to more than one version of Chiron's death, he was accidentally killed by a poisoned arrow that struck his foot. The culprit was Heracles (Hercules), in yet another example of the sheer chaos that seemed to accompany our hero wherever he went [3]. Chiron, the master of healing, could not heal himself from the poison, so Zeus took pity on him and placed him in the stars. This heroic demise gave us the constellation Centaurus, Alpha Centauri, and Chiron's Waterworld, with its future inhabitants likely to be told that they live in a place close to where their namesake was poisoned to death. Actually, the arrow struck Chiron's left foot, which is where the second-brightest star system in Centaurus, Beta Centauri, is situated. This system is much further away than Alpha Centauri, at about 120 parsecs [1], and its three stars are much brighter and younger than either Alpha Centauri or our Sun [42]. The constellation Centaurus and Chiron will both be in our sky for a long time, but not forever, and thousands of years from now Chiron will not look the same: stars will shift their positions in the night sky, due to their different speeds and directions of movement in our Milky Way galaxy [43]. Within only a few hundred thousand years, the movements of the stars will distort Chiron out of all recognition, leaving nothing behind but a legend—provided that there are still any human beings around to remember him.

Proxima Centauri b *How similar to Earth: Slightly larger, but significantly weaker magnetic field and proximity to a flare star may be an issue for long-term survival of its atmosphere. At the very least, it would make its atmosphere less dense than it would otherwise be.*
Plausible planet: Warm super-Earth waterworld with reasonable sized but not thick atmosphere, probably colder than Earth, sunward side having some open water, mostly frozen oceans elsewhere. Surface gravity about 30–40% larger than Earth, rather gloomy apart from occasional flares, very windy.
Best case scenario: Some temperate land regions on day side.
Worst case scenario: Atmosphere blasted away by star, a dreary cold desert.
Cultural connection: Chiron's Waterworld.

*   *   *

GL 628, Wolf 1061, HIP 80824: the names just roll off the tongue, don't they? They all indicate the same star, the very next one outward in the Alpha Centauri sector, and this star is a lot more interesting than the alphabet soup of its designations. From this prolix list of acronyms, we'll pick Wolf 1061 because that is what they call it at exoplanet.eu, but for convenience we will call it Wolfy. A little more than three parsecs away from Alpha Centauri and a little across from it in the nearby constellation of Ophiuchus, Wolfy was first catalogued in the early twentieth century and is an M star like Proxima but somewhat larger and brighter. Unlike Proxima, Wolfy does not seem to be experiencing strong flares. Nevertheless, its radiation output is not entirely stable, being a variable star of the BY Draconis type. This type of variation is modest and is caused by the rotation of the star and the presence of starspots on it that cause changes in its apparent brightness, just like sunspots on our Sun. In the case of Wolfy, it rotates once every 89 days and its typical brightness variations over this time are only about 1–2%, not climatically significant and not a threat to habitability [44].

This is just as well, because in 2015, a team from the University of New South Wales in Australia discovered that Wolfy has three planets [45]. Not a lot is known about them yet. The closest to Wolfy, Wolf 1061 b, is only slightly larger than Earth but receives more stellar radiation than Venus, so is highly unlikely to be habitable (see Table 2.1). The farthest one from Wolfy, Wolf 1061 d, is at least seven times as massive as the Earth and so is probably a sub-Neptune rather than a rocky terrestrial planet. It has a very eccentric orbit that over a 217-day trek takes it as far as 0.73 a.u. from Wolfy and as close as 0.21 a.u. Even at closest approach, though, it is still very cold, with an equilibrium temperature of about −100 °C. As a likely sub-Neptune world, it might have an atmosphere of hydrogen and helium plus perhaps some whitish, persistent clouds of carbon dioxide and water vapor floating above an envelope of water. Like Stevenson's World (see Chap. 1), an atmospheric surface pressure of only a few hundred times that of Earth plus the incoming radiation from Wolfy would likely be enough to keep the ocean surface warm enough to be liquid, although the planet would not be very Earth-like due to its high surface gravity and thick, unbreathable atmosphere.

As for the planet orbiting between these two worlds, Wolf 1061 c, its proximity to its star puts it in hot super-Earth territory, but its equilibrium

Table 2.1 The planets of Wolf 1061, in order of distance from the star

Planet	Mass (Earth = 1)	Orbital period[a] (days)	Distance from star (Earth distance from Sun = 1)	Equilibrium temperature (degrees K) (Earth = 255 K)	Type
Wolf 1061 b	>1.91	4.9	0.038	418–457	Hot super-Earth
Wolf 1061 c	>3.41	17.9	0.0890	271–296	Hot super-Earth
Wolf 1061 d	>7.7	217	0.47	118–129 (176 at closest approach)	Cold sub-Neptune

[a]The period is the time taken to complete one full orbit

temperature is only slightly higher than Earth's. Its actual surface temperature might not be too hot at the surface if its greenhouse effect is modest. Unfortunately, it has more than three times the mass of Earth, and so is much more likely both to have a thicker atmosphere than Earth and more water. Even so, if we assume a thin, Earth-like atmosphere and a global ocean, extrapolation of some previous climate model simulations suggests a day side maximum temperature of about 40 °C (104 °F) and night side minimum temperatures about 20 °C (68 °F) [25]. This could be a tropical, watery world. An analogy could be to Venus, not the real present-day Venus but to the kind of Venus that scientists used to dream about in the days before the Mariner 2 spacecraft flew by Venus in 1962 and confirmed that it was actually hotter than Hades [46]. This pre-Mariner Venus was depicted by science fiction writers as a humid world with thick cloud, lots of rain and thunderstorms, a lukewarm ocean bursting with sea life and with perhaps a few small islands dotted about, inhabited by dinosaurs or other fanciful creatures. Edgar Rice Burroughs wrote a whole series of novels set on this kind of Venus. Heinlein's *Between Planets* (1951) and *Podkayne of Mars* (1963) are also classic examples of this genre [47, 48]. A similar depiction of the planet is contained in Pohl and Kornbluth's influential novel *The Space Merchants* (1952), while an example of the niche market of Christian science fiction, but in the same vein, is C.S. Lewis's *Perelandra* (1943) [49, 50]. The temperature measurements made by Mariner 2 exploded this old Venus into a million evaporating raindrops. But perhaps one day we'll find an extrasolar planet just like the old Venus— who knows, maybe Wolf 1061c could be it.

Actually, the new Venus, the real one, may once have been a little like the old Venus. There is strong evidence that billions of years ago Venus had quite a bit of water, including perhaps surface oceans. Some modelling work suggests that Venus could have been habitable as recently as 700 million years ago [51]. It is possible that since then some kind of major volcanic event dumped a large amount of carbon dioxide into the atmosphere of Venus at a rate so rapid that the extra carbon dioxide couldn't be absorbed by the surface rocks fast enough to stop the atmosphere warming up [52].

The drastic evolution of the climate of Venus, from old Venus to new Venus, highlights an important aspect of human nature: there is a tendency to think of the climates of planets as having been always the same as they are at present. This is not the case for Venus and not even for Earth: as recently as 20,000 years ago, during the peak of the last ice age, there was an ice sheet about 1000 m thick covering the present location of the city of Chicago [53]. On Mars, there is also plenty of evidence of recent ice ages and resulting variations in surface ice content. There is also good geological evidence of lakes and rivers several billion years ago, and possibly even a large Northern Hemisphere ocean at that time [54, 55]. Today almost all surface water on Mars is frozen solid, with only the possibility of random seepage flows here and there when conditions are warm enough [56, 57]. In the case of Wolfy's planets, a possible source of climate change is a change in the brightness of Wolfy itself, as like Proxima, just after its formation Wolfy was brighter than it is now and probably had lots of flares. Since Wolfy is not really a flare star anymore and flares are much more frequent when M stars are young, Wolfy is likely a few billion years old, but this is not known with any accuracy. So early in its evolution the planet Wolf 1061 c was likely even warmer than it is now, making it more difficult for the planet to retain its water and avoid a runaway greenhouse at that time. Nevertheless, since Wolf 1061c is considerably larger than Venus and has a stronger gravity, it would also have been more difficult for water vapor to escape from its atmosphere and it would have been less likely than Venus to lose its oceans. Wolfy may still be like old Venus rather than new Venus, but a runaway greenhouse climate like new Venus remains a distinct possibility.

Wolfy is located in the constellation Ophiuchus. In ancient times, Ophiuchus was depicted in classical Greek mythology as a strong man holding serpents at bay with his bare hands. The constellation has been identified with the Trojan priest Laocoon [58], a legendary figure who reportedly lived during the war between the city of Troy in modern-day western Turkey and a coalition led by the city of Mycenae in central Greece, a conflict known today as the Trojan War. Laocoon did not have much luck during that war. One evening in about 1184 B.C.E., the Greeks besieging Troy had apparently just abandoned their attempt to conquer the city after years of trying and had left in front of the gates of the city a wooden horse, a symbol of Troy, apparently as a parting gift [59]. This gate was most likely the main south gate rather than the smaller east gate that is popular these days with the twenty-first century tourists who pass through the walls of the city [60]. There are varying accounts of what happened to Laocoon, but in one of the most familiar tales, he was suspicious about the wooden horse and said so, mentioning that he didn't much like the idea of Greeks bearing gifts. The god Poseidon then sent two serpents to strangle Laocoon along with his two sons before he could really raise the alarm among his fellow Trojans (Fig. 2.3). It is not entirely clear why the gods killed Laocoon: some legends say it was for doing the right thing in uttering a clear warning, others say it was because this was the wrong thing to do. Then again, determining the motivations of Greek gods is sometimes a useless exercise, like trying to teach empathy to a cat. In any event, the good citizens of Troy decided to bring the horse into their town and to wildly

Fig. 2.3 Laocoon and his Sons, Vatican Museum. Reprinted from Wikipedia (https://en.wikipedia.org/wiki/File:Laoco%C3%B6n_and_his_sons_group.jpg, accessed on the 2nd of July 2024). © 2024, Wilfredo Rafael Rodriguez Hernandez. Public domain

celebrate their apparent victory over the hated enemy. We all know the rest of the story: crouching inside the horse was a small platoon of the best warriors in Greece, including Menelaus, their leader, and Odysseus, later the protagonist of the Odyssey. Once they were wheeled inside the city, they waited until the sounds of partying outside had subsided, then crept out, killed the gate guards and opened the gates to the Greek army hiding near the city. Archaeologists can still see the evidence, some distance below the present-day surface, of the sack of Troy that followed, with human remains scattered among the burnt buildings [61].

It's fun to imagine what might have happened if the Trojans had listened to Laocoon, and if instead of hauling the horse into their city they had rolled it down the hill, thereby smashing it to pieces in the valley below, along with the legendary warriors inside it. The Greeks would then have been thoroughly demoralized: their last stratagem having failed and their leader dead, they would almost certainly have given up the siege and sailed away. But what would have happened afterwards? Would Greek minstrels have treasured for centuries the tale of the attempted siege of Troy and its ignominious end in a

Greek defeat, until the tale could finally be written down? It seems very unlikely; if defeats are to be remembered in song, they usually have to be heroic ones, not embarrassing failures. But if the tale had not been told, then we would never have heard of Laocoon, of Odysseus, or of Troy, and the constellation Ophiuchus would be the subject of a very different story.

Wolf 1061 c *How similar to Earth: Larger than Earth, likely hotter. Possible issue with some atmospheric stripping by flares early in its evolution but the star is no longer a flare star.*
Plausible planet: Tropical super-Earth waterworld with hot oceans and atmosphere thicker than Earth's. Tidally locked but if atmosphere is thick, only moderate temperature difference between day and night side.
Best case scenario: Warm but not hot ocean world, old Venus.
Worst case scenario: Runaway greenhouse, the real Venus.
Cultural connection: Ophiuchus, Laocoon and old Venus.

*   *   *

We move on now to GJ 581, another M star, located this time in the zodiac constellation Libra about 5.5 parsecs away from Earth. Libra is known to have been associated with the idea of scales or a balance for a very long time. It was described as "the scales" or "the balance" in a Babylonian constellation list that begins with the Pleiades, a well-known star cluster in Taurus, full of young, blue stars about 100 million years old and about 135 parsecs away (Chap. 7). It has been speculated that the first constellation on this list is supposed to indicate where the Sun is located at or about the spring equinox (about March 21). However, the Sun was last seen in the Pleiades at this time of year during the Bronze Age around 2000 B.C.E., as nowadays the Sun is located in Aries at the Northern Hemisphere spring equinox. Thus this constellation list may be at least as old as the Bronze Age, and therefore so is the idea of Libra as "the scales". GJ 581 appears to be close in the sky to Beta Librae, otherwise known by its Arabic name Zubeneschamali (see Fig. 2.1) [62]. Actually, Zubeneschamali is a hot B-type star located considerably further away, at about 57 parsecs. It is notable that Eratosthenes, the renowned Hellenistic Greek scientist and geographer who died in about 194 B.C.E., said that Zubeneschamali was brighter than Antares [63]. Since today Antares is very much brighter than Zubeneschamali, and since we have to assume that Eratosthenes was not a complete idiot, either Antares has become considerably brighter in the last 2000 years or Zubeneschamali has become dimmer, or some combination of the two. We just don't know. Zubeneschamali would also be clearly visible in the night sky from any planets of GJ 581, in roughly the same position as in our sky, and being closer to GJ 581 than to us would be slightly brighter than we see it.

Currently, three planets are confirmed circling GJ 581 (Table 2.2), and all of them are likely too hot to be habitable. The planet GJ 581 c is the most potentially habitable of the three but it is still very warm by terrestrial standards, and being at least five times as massive as Earth it is likely a hot sub-Neptune. GJ 581 is an old star, more than 8 billion years old, so unlike many other M stars it is very inactive, and it no longer causes the atmospheric stripping that would have occurred when the star was much younger [64]. A massive planet like GJ 581 c has probably had plenty of time to produce a dense atmosphere of its own, perhaps consisting of steam or of hot carbon dioxide. Simulations indicate that this planet could have suffered a runaway greenhouse, with estimated surface temperatures of a few hundred degrees C [65]. As a result, it is likely inhospitable to terrestrial life.

The above phrase "three planets have been confirmed" highlights one of the difficulties with detecting planets around other stars using the radial velocity method. This method gives raw data on the backwards and forwards movement of GJ 581 in the form of frequency changes in the light emitted by GJ 581. If there were only one planet orbiting GJ 581, it would be relatively easy to confirm its existence, as then if the planet were big enough, the resulting frequency changes would be regular and clear. But if there were many planets, the original radial velocity signal might be the result of

Table 2.2 The planets of GJ 581, in order of distance from the star (luminosity of GJ 581 from [64])

Planet	Mass (Earth = 1)	Orbital period (days)	Distance from star (Earth distance from Sun = 1)	Equilibrium temperature (degrees K) (Earth = 255 K)	Type
GJ 581 e	>1.65	3.15	0.029	505–552	Hot super-Earth
GJ 581 b	>15.19	5.4	0.041	425–464	Hot sub-Jovian
GJ 581 c	>5.65	12.9	0.074	316–346	Hot sub-Neptune
GJ 581 g (unconfirmed)	>2.2	32	0.13	238–261	Warm super-Earth
GJ 581 d (unconfirmed)	>5.6	67	0.218	184–201	Cold sub-Neptune

different possible combinations of planetary orbits and sizes. This source of confusion has led two extra planets to be proposed orbiting GJ 581, namely GJ 581 d and GJ 581 g [66]. If planet g exists (which it may not [67]), it is the most intriguing of the two, with an equilibrium temperature comparable to Earth's and a mass somewhat more than twice that of Earth. The planet would likely be tidally locked, but as we saw in the case of Chiron's Waterworld, this is not an insurmountable obstacle to having terrestrial-type temperatures somewhere on its surface. If we assume a water world with an Earth-type atmosphere, simulations of the climate of GJ 581 g suggest a lobster pattern of Arctic summer temperatures on the day side, like Chiron's Waterworld (see Fig. 2.2). Just like Chiron's Waterworld, climate model simulations where the carbon dioxide concentration was increased to 500 times that of Earth barely budged the day side temperature but did melt all of the ice on the night side [25]. Still, before we get too excited, the planet first needs to exist in order to be studied further.

One aspect of the GJ 581 system that is not in dispute is the presence of a large Kuiper belt in the outer part of the system, extending from about 25 a.u. to about 60 a.u. from the star [68]. Its presence was first deduced in 2012 from infrared images taken by the Herschel Space Observatory of the European Space Agency. The mass of the GJ 581 Kuiper belt appears to be somewhat larger than its solar system equivalent, which means that in the GJ 581 belt there are likely to be hundreds of thousands of objects with diameters of 100 km or greater, with many bigger than 1000 km in diameter.

Before we exit the GJ 581 system, a few more words about our friend Eratosthenes. Our modern advances in astronomy would have surprised him but probably would not have amazed him, as he was the first person to calculate accurately the circumference of the Earth [69]. To describe a simplified version of his technique, he did this by noting that at the Northern Hemisphere summer solstice (June 21) at midday at Syene, a location in Egypt almost due south of Alexandria, the Sun was directly overhead. At Alexandria, though, the Sun was not directly overhead at the same time. He then measured the angle of the Sun to the vertical at Alexandria, combined that with the well-measured distance between Syene and Alexandria, added some simple geometry and the result (perhaps aided by a little luck) was a calculation of the circumference of the Earth that is accurate to a couple of percent [70]. He knew that it was possible in principle to use measurements and geometry to calculate the distance to the Moon, since he knew that the Moon has a noticeable parallax. Indeed, later ancient astronomers like Ptolemy knew this distance, as the great second century B.C.E. astronomer Hipparchus had already performed this calculation [71]. So the idea that the stars also might have a parallax would have been challenging for Eratosthenes but probably not astounding, and neither would have been the idea that they might have planets.

GJ 581 g (Unconfirmed) *How similar to Earth: Only slightly larger and somewhat cooler than Earth. Tidally locked, weak magnetic field likely, potential atmospheric stripping earlier in evolution.*
Plausible planet: Super-Earth that is colder than Earth, similar to Chiron's Waterworld.
Best case scenario: Ocean world, Arctic summer on day side.
Worst case scenario: Thin atmosphere, mostly frozen desert.
Cultural connection: Babylonian Libra.

*   *   *

Located in Scorpius near its tail (see Fig. 2.1) and only 4.5 parsecs from GJ 581, GJ 667 consists of three stars and a bunch of planets, but exactly how many planets remains unclear. Scorpius is one of the few constellations where it is fair to say that its actual shape in the sky clearly resembles what it is supposed to be from its zodiacal name: the claws and long curving sting of a scorpion. As a result, it is one of the most easily spotted constellations, an identification aided by the bright red glowing coal of Antares in a position close to the scorpion's head. This pattern recognition exercise is of course not helpful for ancient peoples who unlike the Babylonians lived in lands that had no scorpions [72]. For instance, the Hawaiians call this constellation Maui's fishhook [73]. Maui is a Polynesian demigod who was credited with using his magic fishhook to pull the Hawaiian Islands from the bottom of the sea. According to the Maori of New Zealand, Maui used a similar trick to create the North Island of that country [74].

So, scorpion or fishhook? Either way, GJ 667 lives near the pointy end. The biggest of its three stars, GJ 667A, is a K star, emitting only about 14% of the Sun's visible radiation. Compared to the Sun, it is also very depleted in heavy elements, that is elements other than hydrogen or helium. It has been shown that stars with a smaller percentage of heavy elements generally form less massive planets [75]. Indeed, no planets have yet been detected orbiting GJ 667A. The second star, GJ 667B, is another K star separated from GJ 667A by an average distance of about 13 a.u., or slightly more than the distance from the Sun to Saturn. The orbit of the stars around each other is quite oval-shaped and combined with their relative closeness to each other might pose some gravitational difficulties for terrestrial planetary formation around both of them [76].

It is the third star that is currently of intense interest. GJ 667C is a decent-sized M star about 230 a.u. away from its companion stars, and it has at least five planets orbiting it [77]. We will call it Hook. Of the planets listed in Table 2.3, we'll focus on two of them, because they are likely to have climates that are most similar to Earth. At first glance, Hook f appears definitely colder than Earth, with an equilibrium temperature only a little higher than that of Mars. It would need a significant greenhouse effect to have a climate as warm as Earth's. Since the planet is larger than Earth, it at least has the potential to have a very thick, carbon-dioxide rich atmosphere with the required greenhouse effect. Better is Hook c, whose equilibrium temperature is similar to Earth's or slightly higher. It is more massive than Hook f, so it really belongs to the warm or hot super-Earth category, with an estimated radius of about 1.5 times that of Earth. As we have already mentioned, a big world like this orbiting around an M star is more likely than not to be a water world, but in a way that is a good thing, because having a world-wide, deep ocean would stabilize the climate in the short term and moreover provide a good location for sea life.

I am careful to say "in the short term" because there are good reasons to believe that the climate of a water world with a deep ocean might be less stable

Table 2.3 The planets of GJ 667C (Hook), in order of distance from the star

Planet	Mass (Earth = 1)	Orbital period (days)	Distance from star (Earth distance from Sun = 1)	Equilibrium temperature (degrees K) (Earth = 255 K)	Type
GJ 667C b	>5.59	7.2	0.05	392–429	Hot super-Earth or hot sub-Neptune
GJ 667C c	>3.8	28.1	0.125	249–273	Warm super-Earth
GJ 667C f	>2.7	39.0	0.156	223–244	Warm super-Earth
GJ 667C e	>2.7	62.2	0.213	191–209	Warm or cold super-Earth
GJ 667C d (unconfirmed)	>5.1	91.61	0.276	168–183	Cold super-Earth
GJ 667C g	>4.6	256.2	0.549	119–130	Cold super-Earth

over periods of tens or hundreds of millions of years than the climate of a world that has continents. This is due to something called the inorganic carbon cycle, described particularly well in a 2014 paper by Yann Alibert from the University of Bern [78]. Carbon dioxide is routinely removed from the atmosphere by rainfall and the carbon dioxide in the rain then reacts with silicate rocks, turning them into carbonates, a process known as silicate weathering (Fig. 2.4). The carbonates then flow into the ocean and sink to the bottom. Then, over millions of years, the seafloor moves due to continental drift, or "plate tectonics" as it is more properly termed [79]. Plate tectonic movement takes the carbonates beneath the continents, a process known as subduction that thereby removes the carbon dioxide from the atmosphere-ocean system. Volcanoes gradually return the carbon dioxide to the atmosphere or can add additional carbon dioxide. On a planet with life, organisms in the ocean can also use the carbonates to create their shells, and when they die, their shells fall to the bottom of the ocean where they can also then be subducted. This is known as the organic carbon cycle. The reason that the carbon cycle can stabilize the climate is that if the temperature decreases, the rate of silicate weathering also decreases due to lower evaporation and therefore lower rainfall, but volcanoes still release carbon dioxide into the atmosphere at the same rate.

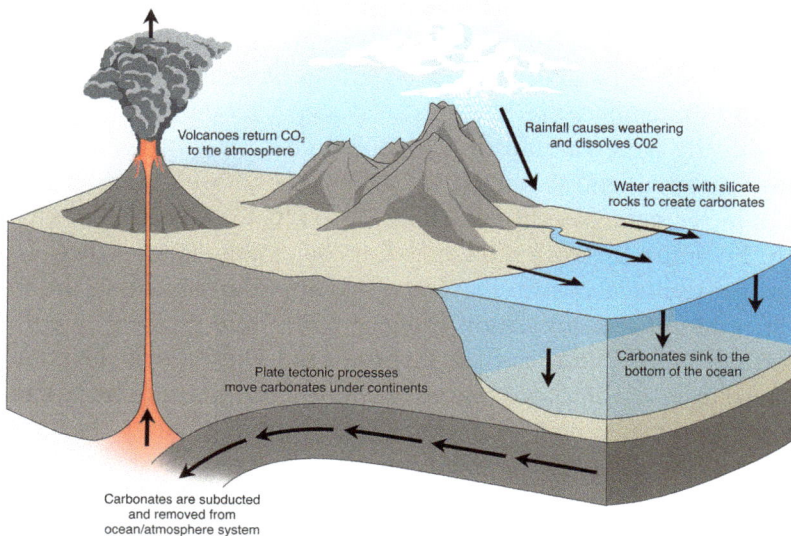

Fig. 2.4 The carbonate-silicate cycle. Printed with permission. © 2021, Kevin Walsh, Gemma van Hurk. All rights reserved

Eventually, this will increase the carbon dioxide content of the atmosphere and the size of the greenhouse effect, leading to an increase in temperature and thus counteracting the original temperature decrease. On the other hand, let's say the temperature initially increased for some reason. This would increase the evaporation of water, the rainfall rate and also the rate of silicate weathering, thus removing carbon dioxide from the atmosphere. This would decrease the greenhouse effect and thereby the planetary temperature, so counteracting the original temperature increase.

For this stabilizing cycle to work, there are a couple of prerequisites. First, the bottom of the planet's ocean has to be in contact with the rocks below it. If not, there would be no way for the carbonates to be carried away by sub-duction as the continents move around. This is problematic for planets with a deep ocean, say more than about sixty kilometers, as the water pressure at the bottom of such an ocean would be so immense that the water there would be compressed into ice [80]. This ice layer might isolate the ocean from the plate tectonic processes, wrecking the carbon cycle at the same time. Second, there needs to be a significant internal heat source to keep the interior of the planet molten, so that the plates can still slide around and so subduction can occur. Older and smaller planets are less likely to have the required internal heat, thus stopping plate tectonics and subduction.

Whether or not the planet Hook c experiences a carbon cycle, it likely has a slow rotation, with calculations suggesting that due to its eccentric orbit the most likely rotation state is a 3:2 resonance [81]. This means a sol of about 56 Earth days or twice the orbital period. The maximum temperature in the equatorial region is naturally almost certain to be considerably higher than the equilibrium temperature, which is an average over the entire planet. For a planet with the same amount of greenhouse effect as Earth, the maximum daytime equatorial land temperature could easily be 50 K warmer than the equilibrium temperature, thus reaching 296–320 K, which is 23–47 °C or 73–115 °F—warm by Earth standards. But cooler temperatures there are also possible. For technical reasons that I won't go into, the planet is in the "slowly rotating" atmospheric circulation category [82], here typified by rising air near the sub-stellar point, causing intense cloud and rainfall there. Surface winds head from all directions towards the sub-stellar point to replace the ris-ing air, in a giant monsoon wind of the kind that brings rain to India during the summer (Fig. 2.5). The outcome would be a region near the sub-stellar point full of showers and thunderstorms that would then migrate around the equator following the sun, thus cooling this region considerably [83]. Conversely, nighttime would be a period of much drier and calmer weather. A slight modification to this picture is that the rotation of the planet would generate a lobster pattern, shifting the region of maximum rainfall a couple of thousand kilometers east of the sub-stellar point, but the basic idea is the same.

When the skies above the surface of Hook c are clear, an unusual feature will be the presence of two extraordinarily brilliant stars. GJ 667A and GJ 667B would appear as a couple of sharp orangey points of light each roughly about as bright as the full Moon as seen in our sky, and they would be easily visible during daytime. Their apparent distance from each other in the sky of Hook c would vary but could be as much as about four times the width of Earth's Moon in our sky. Depending on the orbit of this planet around its star, though, they would not be visible from all locations on the planet at all times, and in fact there might be some latitudes on the planet where they might not be seen, just as Alpha Centauri cannot be seen from New York.

Another aspect of the environment of Hook c needs to be mentioned. The orbit of Hook c around its star is slightly eccentric, so it experiences a significant amount of internal tidal heating. Estimates suggest that this is at least a thousand times the tidal heating that affects Earth [84]. A back-of-the-envelope calculation indicates that tidal heating on Hook c would then be larger than the internal heating on the planet caused by radioactive decay or residual heat from its formation, if the planet had a similar inventory of radioactive elements to Earth. As a result, this suggests a world that is somewhat more volcanic than Earth, all other things being equal. We are not talking about a planet full of geysers and fumaroles, like the most volcanic areas of Iceland or Yellowstone Park. That would require roughly an average of at least a few thousand times Earth's tidal heating, or at least a few hundred times as much as the total internal heating reaching the surface of the Earth [85]. Even so, given how inaccurate these estimates are, this scenario is still not out of the question for Hook c. This amount of internal heating pales into insignificance, though, when compared to that on the innermost planet in the system, Hook b. Tidal heating on this world is calculated to be more than 100,000

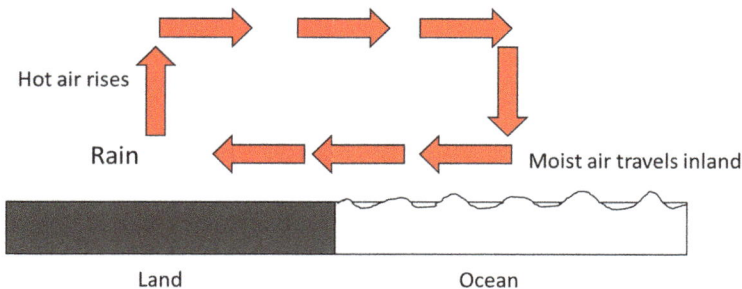

Fig. 2.5 Schematic diagram illustrating the monsoon process

times greater than Earth's, which gives a typical surface much more volcanically active than Yellowstone. This suggests that just like Io, Jupiter's moon,
the innermost planet's surface could be bursting with geysers and volcanoes
and may even be completely molten [84].

GJ 667C c *How similar to Earth: A largish super-Earth, likely tidally locked in
3:2 spin orbit resonance. Located near the inner edge of the conservative habitable
zone, rather weak magnetic field likely. Likely more volcanic than Earth.*
Plausible planet: A hot, watery world with a thick atmosphere.
*Best case scenario: Hot but not necessarily uninhabitable conditions on the day
    side, milder conditions on the night side. Atmosphere not crushing.*
Worst case scenario: Super Venus. Dry, oven-baked world.
Cultural connection: Maui's fishhook.

*   *   *

The final planetary system in our tour of this sector is that of the isolated,
Sun-like star 61 Virginis [86]. As the name suggests, it is in Virgo, and is just
bright enough to be seen by an experienced observer with the naked eye. It is
both slightly cooler and slightly older than the Sun but has almost twice its
radius, consistent with our understanding of how stars like the Sun will evolve
in the future, as like humans they tend to expand as they age. Its system
includes a big super-Earth as well as a couple of Neptune-sized planets, all of
which are far too hot to be habitable. It also has a giant Kuiper-like belt with
perhaps as many as ten times the number of objects in it as in our Solar
System's Kuiper belt. Curiously, the belt of 61 Virginis appears to be clear
inside of 30 a.u. from the star [87]. This strongly suggests that there are additional undiscovered planets orbiting in this region that have swept it free of
asteroids and comets. Also, planets the size of Earth orbiting at habitable
distances from the star have not been ruled out at present but have not been
detected either [88].

The 61 Virginis system is reliably estimated to be at least 6 billion years old
[89]. At this age, any Earth-sized planets in the system's habitable zone would
have been in existence at least 1.2 billion years longer than our own Earth.
This extra time could have given life a better opportunity to develop but
because the star is evolving, its brightness is gradually increasing. Thus the
boundaries of its habitable zone have expanded outwards over the billions of
years since its formation. The resulting higher temperatures on a habitable
planet in this system would have greatly affected its ecology. Some clever work
by British astronomers led by Jack O'Malley-James (now at Cornell) has made

some informed speculation about the kind of biosphere that might exist on such an old planet [90]. They have made this deduction from a comparison to the past and projected future evolution of the biosphere of Earth itself. Present-day climate conditions on Earth, where our world hosts a diverse menagerie of multicellular organisms, are very unlikely to still exist after about 1 billion years from now. The increasing brightness of our Sun will lead to rising temperatures and a resulting die-off in vegetation, along with declining oxygen levels in our atmosphere. The most likely kind of biosphere today for an old habitable zone planet orbiting 61 Virginis is described by O'Malley-James as "Microbial (declining)", characterised by high temperatures, an absence of oxygen, and the presence of simple life in the form of microbes — just like our poor old Earth when it gets to that age.

Further Afield

We will talk a lot in this book about constellations made up of stars. Less well recognized is that some myths describe features that are made up of the dark spaces in the sky. The First Nations people of Australia describe the Emu in the Sky, comprising the dusty bits of the Milky Way between the Southern Cross and Altair, or along the shaded bit of Fig. 2.1 and the lower right part of Fig. 3.1. We could call this object a "nonstellation" [91]. Its appearance changes with the seasons, corresponding to the seasonal behavior of the giant bird and its relationship to its environment.

Just outside our 10 parsec limit in this sector is BPM 37093, a very dense white dwarf star that has about the same mass as the Sun but with diameter of only about 4000 km, or less than half that of Earth. It is made mostly of crystallized carbon, otherwise known as diamond [92]. Just knowing this snippet of information is highly unlikely to start a new diamond rush, however: even allowing for the fact that we do not yet have a practical means of interstellar flight, the surface temperature of the star is over 10,000 K, making mining operations there a very challenging proposition. Just across the northern boundary of this sector, in the Arcturus sector (Chap. 4) at 18 parsecs distant is 70 Virginis b, one of the first extrasolar planets to be discovered [93]. This hot super-Jovian has an eccentric, 117-day orbit around its aging G star 70 Virginis, like 61 Virginis now heading towards red giant status. It is not known whether this system hosts any habitable planets or not, but if it does, they will likely be ancient and probably worn out.

A landmark in this sector is the two B stars that make up Spica, 77 parsecs away and one of the 20 brightest stars in the night sky [94]. They are very young, probably only about 10 million years old, and like all fast-evolving B stars, they will be gone from the sky practically before we know it, in a few tens of millions of years. Further out is Chiron's left foot, Beta Centauri, consisting of three B stars together in the same system, two of which have already started to become giants. Finally, at 170 parsecs is the Antares system, but not for long [95]. Antares will burn bright and swift and will be a dim cinder long before our Sun burns out, and so will Beta Centauri, Chiron's bane; and with the flameout of Beta Centauri, Chiron will figuratively have the last laugh.

Alpha Centauri Sector Summary

The sector that we have called Chiron's Place has plenty to explore: the closest star to Earth, Proxima, along with its super-Earth planet; the diverse, multi-planet system of Wolf 1061; GJ 581 g, the Earth-like planet that may or may not exist; GJ 667, the six-planet stellar system at the pointy end of Maui's fish-hook; and the old 61 Virginis system, where any habitable planets would be on the way out. Naturally, none of these systems could be described as even reasonably well observed. But there are many more stars in this sector than the few that have been described here: if we assume that the density of stars in Chiron's Place is about the same as the average near the Sun, in this sector there would be about 50–100 stars within 10 parsecs, with many undiscovered planets likely. And there are many, many more systems further out than 10 parsecs. It would be tempting to say that our exploration of this region to this point has only scratched the surface, but frankly that would be too generous. One day, though, it is possible that a tiny spacecraft, after a journey of some decades, may reach the Alpha Centauri system, point a camera back at our Sun and transmit the resulting picture back to Earth. As Stephen Dole pointed out many years ago in his remarkable early work on habitable exoplanets, *Habitable Planets for Man* (1964), this image will show an extra star in the constellation Cassiopeia, a bright one that will form an extra bend in the "W" of existing bright stars in that constellation (Fig. 2.6) [96]. This star will be our Sun, and as a result, rather than humanity currently being separated into the various signs of the zodiac depending on what time of year we are born, instead we Earthlings will all become Cassiopeians.

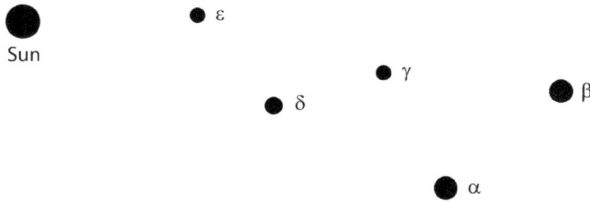

Fig. 2.6 The extra star in Cassiopeia. Modified with permission from [97]. © 1964, The RAND Corporation. All rights reserved

References

1. Van Leeuwen F (2007) Validation of the new Hipparcos reduction. Astronom Astrophys. 474(2):653–664. https://doi.org/10.1051/0004-6361:20078357
2. Ovid (8 B.C.E.). Fasti, 5.379-414. https://www.poetryintranslation.com/PITBR/Latin/Fastihome.php. Accessed 22 Nov 2023
3. Ridpath I (2018) Star tales. Lutterworth, Cambridge, p 69
4. Ptolemaeus C (1984) Ptolemy's Almagest. Translated by Toomer GJ. Gerald Duckworth, London, p 368. Note 136
5. Pannekoek A (1989) A history of astronomy. Dover, Mineol, pp 345–346
6. Rao NK, Vagiswar A, Louis C (1984) Father J. Richaud and early telescope observations in India. Bull Astronom Soc India 12(1):81
7. Kochhar RK (1991) French astronomers in India during the 17th-19th centuries. J Br Astronom Assoc 101(2):95–100
8. Innes RTA (1915) A faint star of large proper motion. Circ Union Obs Johannesburg 30:235–236
9. Anglada-Escudé G et al (2016) A terrestrial planet candidate in a temperate orbit around Proxima Centauri. Nature 536(7617):437–440. https://doi.org/10.1038/nature19106
10. Mascareño AS, Faria JP, Figueira P et al (2020) Revisiting Proxima with Espresso. Astronom Astrophys 639:A77
11. Bixel A, Apai D (2017) Probabilistic constraints on the mass and composition of Proxima b. Astrophys J Lett 836(2):L31
12. Kervella P et al (2020) Orbital inclination and mass of the exoplanet candidate Proxima c. Astronom Astrophys 635:L14. https://doi.org/10.1051/0004-6361/202037551
13. Pineda JS, Youngblood A, France K (2021) The M-dwarf ultraviolet spectroscopic sample. I. Determining stellar parameters for field stars. Astrophys J 918(1):23
14. Ribas I, Bolmont E, Selsis F et al (2016) The habitability of Proxima Centauri b. I. Irradiation, rotation and volatile inventory from formation to the present. Astronom Astrophys 596:A111

15. Alibert Y, Benz W (2017) Formation and composition of planets around very low mass stars. Astronom Astrophys 598:L5
16. Miguel Y, Cridland A, Ormel CW et al (2020) Diverse outcomes of planet formation and composition around low-mass stars and brown dwarfs. Mon Not Roy Astronom Soc 491(2):1998–2009
17. Makarov VV (2013) Why is the Moon synchronously rotating? Mon Not Roy Astronom Soc Lett 434(1):L21–L25
18. Stevenson DJ (1987) Origin of the moon—the collision hypothesis. Annu Rev Earth Plan Sci 15(1):271–315. https://doi.org/10.1146/annurev. ea.15.050187.001415
19. Canup RM (2012) Forming a Moon with an Earth-like composition via a giant impact. Science 338(6110):1052–1055
20. Perkins S (2016) Ancient eclipses show Earth's rotation is slowing. science.com, 2 Dec 2016. https://www.sciencemag.org/news/2016/12/ancient-eclipses-show-earth-s-rotation-slowing. https://doi.org/10.1126/science.aal0469. Accessed 21 Feb 2021
21. Starr M (2020) Ancient shell has revealed exactly how much shorter days were 70 million years ago. Science Alert, 10 Mar 2020. https://www.sciencealert.com/old-shell-reveals-earth-s-days-were-half-an-hour-shorter-70-million-years-ago. Accessed 22 Feb 2021
22. Wolf ET, Kopparapu R, Haqq-Misra J, Fauchez TJ (2022) ExoCAM: a 3D climate model for exoplanet atmospheres. Plan Sci J 3(1):7
23. Del Genio AD, Way MJ, Amundsen DS et al (2019) Habitable climate scenarios for Proxima Centauri b with a dynamic ocean. Astrobiology 19(1):99–125
24. Matsuno T (1966) Quasi-geostrophic motions in the equatorial area. J Meteorol Soc Jpn Ser II 44(1):25–43
25. Hu Y, Yang J (2014) Role of ocean heat transport in climates of tidally locked exoplanets around M dwarf stars. Proc Natl Acad Sci USA 111(2):629–634
26. Ibid. (ref. [23])
27. Damasso M, Del Sordo F, Anglada-Escudé G et al (2020) A low-mass planet candidate orbiting Proxima Centauri at a distance of 1.5 au. Sci Adv 6(3):eaax7467
28. Allred JC, Kowalski AF, Carlsson M (2015) A unified computational model for solar and stellar flares. Astrophys J 809(1):104
29. Charbonneau P, Sokoloff D (2023) Evolution of solar and stellar dynamo theory. Space Sci Rev 219(5):35
30. Howard WS, Tilley MA, Corbett H (2018) The first naked-eye superflare detected from Proxima Centauri. Astrophys J Lett 860(2):L30. https://doi.org/10.3847/2041-8213/aacaf3
31. Parke Lloyd RO (2021) Massive flare on Proxima Centauri could spell bad news for any alien life. astronomy.com (online). https://astronomy.com/news/2021/05/massive-flare-seen-on-the-closest-star-to-the-solar-system-what-it-means-for-chances-of-alien-neighbors. Accessed 8 May 2021

32. NASA (2001) Ultraviolet radiation: how it affects life on Earth. https://earthob-servatory.nasa.gov/features/UVB/uvb_radiation3.php. Accessed 3 May 2021

33. Turbet M, Leconte J, Selsis F, Bolmont E et al (2016) The habitability of Proxima Centauri b II. Possible climates and observability. Astronom Astrophys 596:A112. https://doi.org/10.1051/0004-6361/201629577

34. Garcia-Sage K, Glocer A, Drake JJ, Gronoff G, Cohen O (2017) On the magnetic protection of the atmosphere of Proxima Centauri b. Astrophys J Lett 844(1):L13

35. Niven L (1988) The Man-Kzin wars. Baen, Riverdale

36. Avatar (2009) imdb.com. https://www.imdb.com/title/tt0499549/?ref_=fn_al_tt_1. Accessed 5 Jan 2023

37. Baxter S (2012) The science of Avatar. Gollancz, London, p 118

38. Baxter S (2012) The science of Avatar. Gollancz, London, p 90

39. Wagner K, Boehle A, Pathak P et al (2021) Imaging low-mass planets within the habitable zone of α Centauri. Nat Commun 12(1):922. https://doi.org/10.1038/s41467-021-21176-6

40. Gratton R (2020) Searching for the near-infrared counterpart of Proxima c using multi-epoch high-contrast SPHERE data at VLT. Astronom Astrophys 638:A120

41. Faria JP, Suárez Mascareño A et al (2022) A candidate short-period sub-Earth orbiting Proxima Centauri. Astronom Astrophys 658:17. https://doi.org/10.1051/0004-6361/20214233

42. Ausseloos M, Aerts C, Lefever K et al (2006) High-precision elements of double-lined spectroscopic binaries from combined interferometry and spectroscopy. Application to the β Cephei star β Centauri. Astronom Astrophys 455(1):259–269. https://doi.org/10.1051/0004-6361:2006482

43. Klesman A (2017) Can you imagine the sky in five million years? astronomy.com. https://www.astronomy.com/science/can-you-imagine-the-sky-in-five-million-years/. Accessed 20 Apr 2021

44. Kane SR, von Braun K, Henry GW et al (2017) Characterization of the Wolf 1061 planetary system. Astrophys J 835(2):200

45. Wright DJ, Wittenmyer RA, Tinney CG et al (2016) Three planets orbiting Wolf 1061. Astrophys J Lett 817(2):L20

46. Sonett CP (1963) A summary review of the scientific findings of the Mariner Venus mission. Space Sci Rev 2(6):751–777. https://doi.org/10.1007/BF00208814

47. Heinlein RA (1951) Between planets. Charles Scribner's Sons, New York

48. Heinlein RA (1963) Podkayne of Mars. Putnam, New York

49. Pohl F, Kornbluth CM (1953) The space merchants. Ballantine, New York

50. Lewis CS (1943) Perelandra. Bodley Head, London

51. Way MJ, Del Genio AD, Kiang NY et al (2016) Was Venus the first habitable world of our solar system? Geophys Res Lett 43:8376–8383

52. Way MJ, Del Genio AD (2020) Venusian habitable climate scenarios: Modeling Venus through time and applications to slowly rotating Venus-like exoplanets. J Geophys Res Planets 125(5):e2019JE006276

53. Bromwich DH, Toracinta ER, Wei H et al (2004) Polar MM5 simulations of the winter climate of the Laurentide Ice Sheet at the LGM. J Climate 17(17):3415–3433

54. Wordsworth RD (2016) The climate of early Mars. Annu Rev Earth Plan Sci 44:381–408

55. Kite ES (2019) Geologic constraints on early Mars climate. Space Sci Rev 215:1–47

56. Malin MC, Edgett KS (2000) Evidence for recent groundwater seepage and surface runoff on Mars. Science 288(5475):2330–2335

57. Dundas CM, Becerra P, Byrne S et al (2021) Active Mars: A dynamic world. J Geophys Res Planets 126(8):e2021JE006876

58. Thompson R, Thompson BF (2007) Illustrated guide to astronomical wonders: From novice to master observer. O'Reilly, Sebastopol, CA, p 326

59. Wikipedia (2023) The Trojan War. Accessed 14 Apr 2021

60. Archaeological Institute of America (2021) Uncovering Troy. https://www.archaeology.org/travel/interactivemap-troy/. Accessed 14 Apr 2021

61. Montesanto M (2018) Fall of Troy: the legend and the facts. The Conversation. https://theconversation.com/fall-of-troy-the-legend-and-the-facts-92625

62. Kunitzsch P, Smart T (2006) A dictionary of modern star names: a short guide to 254 star names and their derivations (2nd rev. ed.). Sky, Cambridge, MA

63. Kaler JB (2002) The hundred greatest stars. Copernicus, New York, p 201

64. Bonfils X, Forveille T, Delfosse X et al (2005) The HARPS search for southern extra-solar planets-VI. A Neptune-mass planet around the nearby M dwarf Gl 581. Astronom Astrophys 443(3):L15–L18

65. Leconte J, Forget F, Charnay B et al (2013) 3D climate modeling of close-in land planets: circulation patterns, climate moist bistability, and habitability. Astronom Astrophys 554:A69

66. Cuntz M, Engle SG, Guinan EF (2024) The once-canceled habitable-zone Super-Earth Gliese 581 d might indeed exist! Res Not AAS 8(1):20

67. Dodson-Robinson SE, Delgado VR, Harrell J, Haley CL (2022) Magnitude-squared coherence: a powerful tool for disentangling Doppler planet discoveries from stellar activity. Astronom J 163(4):169

68. Lestrade JF, Matthews BC, Sibthorpe B et al (2012) A debris disk around the planet hosting M-star GJ 581 spatially resolved with Herschel. Astronom Astrophys 548:A86

69. Russo L (2004) The forgotten revolution. Springer, Berlin, p 68. https://doi.org/10.1007/978-3-642-18904-3

70. Russo L (2004) The forgotten revolution. Springer, Berlin, p 273. https://doi.org/10.1007/978-3-642-18904-3

71. Ptolemy C (2014) Donahue WH (ed) The Almagest: Introduction to the mathematics of the heavens (trans: Perry BM). Green Lion, Santa Fe, NM

72. White G (2008) Babylonian star-lore. Solaria, London, p 175
73. Westervelt WD (1910) Legends of Maui—A demi god of Polynesia and of his mother Hina. Hawaiian Gazette, Honolulu, HI, pp 12–31
74. Tregear E (1891) The Maori-Polynesian comparative dictionary. Lyon and Blair, Wellington, p 234. https://archive.org/details/maoripolynesian01treggoog/page/234/mode/2up?q=Maui. Accessed 7 Feb 2024
75. Buchhave LA, Latham DW, Johansen A et al (2012) An abundance of small exoplanets around stars with a wide range of metallicities. Nature 486(7403):375–377
76. Söderhjelm S (1999) Visual binary orbits and masses post HIPPARCOS. Astronom Astrophys 341:121–140
77. L'Observatoire de Paris (2021) Encyclopaedia of exoplanetary systems. http://exoplanet.eu/catalog/. Accessed 29 Mar 2021
78. Alibert Y (2014) On the radius of habitable planets. Astronom Astrophys 561:A41
79. Palin RM, Santosh M (2021) Plate tectonics: What, where, why, and when? Gondwana Res 100:3–24
80. D'Angelo G, Bodenheimer P (2016) In situ and ex situ formation models of Kepler 11 planets. Astrophys J 828(1):33
81. Wang Y, Tian F, Hu Y (2014) Climate patterns of habitable exoplanets in eccentric orbits around M dwarfs. Astrophys J Lett 791(1):L12
82. Del Genio AD, Suozzo RJ (1987) A comparative study of rapidly and slowly rotating dynamical regimes in a terrestrial general circulation model. J Atmos Sci 44(6):973–986
83. He F, Merrelli A, L'Ecuyer TS, Turnbull MC (2022) Climate outcomes of Earth-similar worlds as a function of obliquity and rotation rate. Astrophys J 933(1):62
84. Makarov VV, Berghea C (2013) Dynamical evolution and spin-orbit resonances of potentially habitable exoplanets. The case of GJ 667C. Astrophys J 780(2):124. https://doi.org/10.1088/0004-637X/780/2/124
85. United States Geological Survey (2021) Questions about heat flow and geothermal energy at Yellowstone. https://www.usgs.gov/volcanoes/yellowstone/questions-about-heat-flow-and-geothermal-energy-yellowstone. Accessed 3 Apr 2021
86. Gray RO et al (2003) Contributions to the Nearby Stars (NStars) project: spectroscopy of stars earlier than M0 within 40 parsecs: the northern sample. I. Astronom J 126(4):2048–2059. https://doi.org/10.1086/378365
87. Wyatt MC et al (2012) Herschel imaging of 61 Vir: implications for the prevalence of debris in low-mass planetary systems. Mon Not Roy Astronom Soc 424(2):1206–1223
88. Kopparapu RK, Barnes R (2010) Stability analysis of single-planet systems and their habitable zones. Astrophys J 716(2):1336
89. Mamajek EE, Hillenbrand LA (2008) Improved age estimation for solar-type dwarfs using activity-rotation diagnostics. Astrophys J 687(2):1264–1293. https://doi.org/10.1086/591785

90. O'Malley-James JT, Greaves JS, Raven JA, Cockell CS (2015) In search of future Earths: assessing the possibility of finding Earth analogues in the later stages of their habitable lifetimes. Astrobiology 15(5):400–411

91. Hamacher J (2019) Kindred skies: ancient Greeks and Aboriginal Australians saw constellations in common. The Conversation. https://theconversation.com/kindred-skies-ancient-greeks-and-aboriginal-australians-saw-constellations-in-common-74850. Accessed 12 Nov 2023

92. Metcalfe TS, Montgomery MH, Kanaan A (2004) Testing white dwarf crystallization theory with asteroseismology of the massive pulsating DA star BPM 37093. Astrophys J 605(2):L133

93. Marcy GW, Butler RP (1996) A planetary companion to 70 Virginis. Astrophys J 464(2):L147

94. Wikipedia (2023) Spica. Accessed 1 Nov 2023

95. Wikipedia (2023) Antares. Accessed 1 Nov 2023

96. Dole S (1964) Habitable planets for man. Blaisdell, New York

97. Dole S (2007) Habitable planets for man (reprinted edition). RAND, Santa Monica, CA. https://www.rand.org/pubs/commercial_books/CB179-1.html

3

The Summer Triangle
The Vega Sector: 18–24 h, 0 to +90°

The summer triangle is an easily identifiable three-sided group of stars in the northern summer sky, consisting of the bright stars Vega, Altair and Deneb (see Fig. 3.1). Vega and Altair are both main sequence A stars within 10 parsecs of our solar system, while Deneb is a very luminous A-type supergiant much further out, at about 800 parsecs. The three stars play a prominent role in one of the more romantic tales told about the starry sky, namely the love story of Qixi.

Qixi means "seventh night", and this tale has been told in China since at least the Han dynasty around 200 B.C.E. and is likely much older than that [1, 2]. In the Qixi myth, the beautiful immortal Zhinu (the star Vega) was a weaver and Nuilang (the star Altair) was a simple mortal cowherd. Alas, their love was star-crossed (literally) and Zhinu's family forbade the two of them to be together, exiling them to different sides of a river, represented by the Milky Way. One day a year, though, a bridge across the river is formed by a flock of magpies, represented by the star Deneb, and the lovers are permitted to meet. This meeting takes place in August and today it is celebrated as the Chinese equivalent of Valentine's Day. The word Queqiao, or "magpie bridge", is also the name of a Chinese satellite launched into orbit around the Moon in 2018, so called because it acts as a communications link by relaying signals from Earth to other spacecraft that have landed on the far side of the Moon [3].

The western mythological concept of Vega is rather less romantic. Vega is the brightest star in the constellation Lyra, and the translation of the Arabic name for Lyra is "falling vulture", with "Vega" coming from the part of the name that means "falling" [4]. More life-affirming is the Greek tradition,

K. J. E. Walsh, *Planets of the Known Galaxy*, Science and Fiction,
https://doi.org/10.1007/978-3-031-68218-6_3

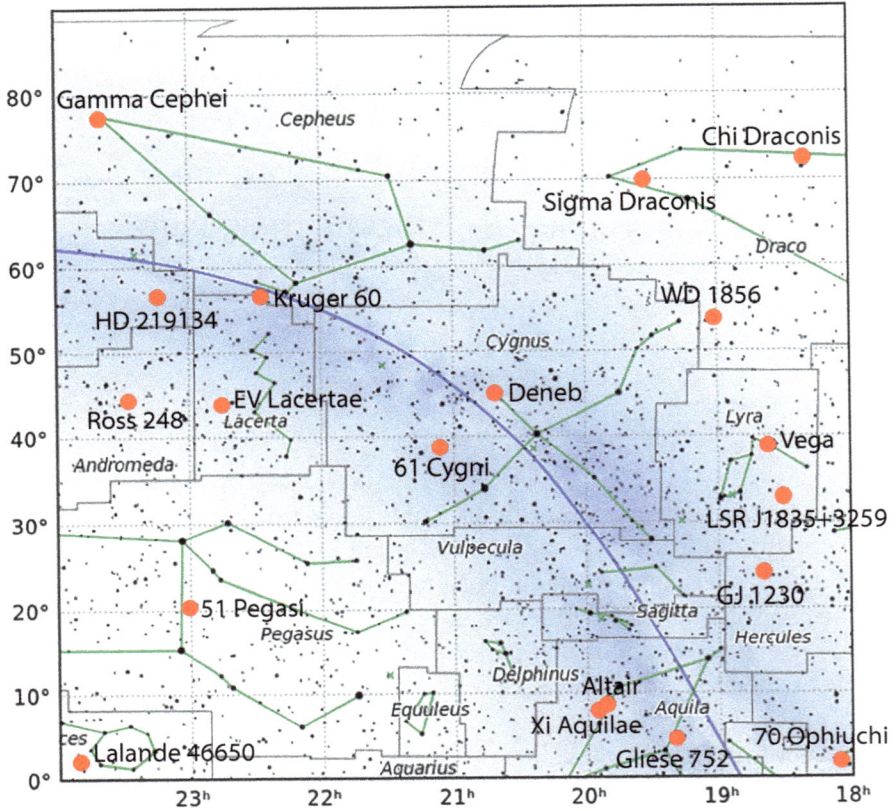

Fig. 3.1 Map of the Vega sector. Blue shading indicates the "Milky Way", the most star-packed part of the galaxy, while the blue (darker) line is the galactic equator, an imaginary circle running through the Milky Way that slices the galaxy in half. The orange (lighter) line is the path of the Sun's location through the seasons (indicating the "zodiac" constellations). Larger circles indicate brighter stars, while locations mentioned in this chapter are indicated by red circles. Image produced using StarCharter (https://github.com/dcf21/star-charter, accessed on the 2nd of July 2024. © 2020, Dominic Ford) software under the GNU general public license

where Lyra is the harp of Orpheus, Greek hero and musician extraordinaire, whose doomed love for Eurydice has been the topic of numerous works of art and even of a comic opera by Offenbach (*Orpheus in the Underworld*). Vega has about twice the mass of our Sun and compared to the Sun is quite egg-shaped, due to its fast rotation that causes the equatorial regions, the fastest moving latitudes, to bulge outwards. It rotates in only about 15 h compared to about 25 days for the Sun. In fact, Vega is rotating so rapidly that it would only have to spin about 20% faster for it to start breaking apart due to excessive centrifugal force at its equator [5].

Since Vega is a bright, nearby star, it has been portrayed endlessly in science fiction. It was noted in passing in Asimov's *Foundation* series (1951) [6] and Vegans were major villains in James Blish's *Cities in Flight* (1955–1962) [7]. In Heinlein's juvenile novel *Have Space Suit, Will Travel* (1958) [8], the Vegans are much nicer, as the fifth planet of Vega, Vega V, is home to the peaceful, capable and nurturing Mother Thing people. The fictional six planets of Vega are important locations in Jack Vance's *Demon Princes* series (1964–1981) [9], while in Carl Sagan's only science fiction novel, *Contact* (1985) [10], radio signals coming from the vicinity of Vega are an integral part of the plot.

Vega is only 455 million years old [5]. When the Earth was the same age, a little over 4 billion years ago, it had some surface water but little or no life. While we have no knowledge at present regarding how fast or whether life might emerge on any hypothetical planets of Vega, by analogy to the development of life on Earth it would be easy to argue that any life in the Vegan system would be at an early, likely microscopic stage of development. Planets circling Vega that host sophisticated macroscopic life, full of jungles and vicious predators, cannot be ruled out at this stage but seem unlikely.

Vega is also about 45 times brighter than the Sun, meaning that for a planet orbiting it to receive about the same amount of stellar radiation as the Earth, the planet would have to orbit at about 7 a.u. from Vega. This is not a real problem as there appear to be no theoretical reasons why there cannot be terrestrial planets orbiting Vega at this distance. Nevertheless, as an A0 star on the border between the A and B stars, Vega has a much hotter surface than the Sun, and so its peak amount of outgoing radiation is in the UV region of the spectrum rather than the visible part. Thus any planets at 7 a.u. from it will be bombarded with intense amounts of damaging UV. For example, in Heinlein's *Have Space Suit, Will Travel,* it is correctly noted that the level of UV on Vega V is unhealthy for human beings. For such a planet to have habitable land regions, it will need a very strong protective ozone layer to block the UV, and even that might not be sufficient to prevent dangerous amounts of UV striking the surface. This situation is not unprecedented because Earth would have been a little like this billions of years ago when Earth was as old as Vega. At that time, Earth would not have had a protective ozone layer, because ozone requires oxygen and there wasn't much on Earth back then. Even given that the Sun produces much less UV than Vega, in those days the absence of an ozone layer would still have permitted enough UV to reach Earth's land surface to cause early lifeforms significant damage [11].

As mentioned earlier, this problem would not be so relevant to ocean life, as a depth of some tens of meters of water provides an effective UV shield. Also, early life forms on Earth did not need oxygen to survive. For some of them, like the anaerobic bacteria that live in the human digestive tract, oxygen is poisonous [12]. For them, the lack of oxygen is a bonus, and they could quite happily live under the sea of a watery Vegan planet, with no competition from nasty, oxygen-hungry species. But they would be generally low-energy organisms. The huge advantage of organisms that use oxygen, like plants, is that they produce energy more quickly [13]. Even more energy is available to organisms that can breathe in large amounts of oxygen. This explains why cheetahs can move faster than, say, mango trees—and much faster than anaerobic organisms.

For any such creatures to survive on the land regions of a Vegan planet, they would need protection from UV, either through evolution of some kind of shield as part of the organism itself, or from an ozone layer. On Earth, the oxygen required for an ozone layer was supplied by photosynthesis, but substantial photosynthesis did not start on Earth until it was about a billion years older than Vega is now [14]. But that is not the only way to obtain an atmosphere with substantial oxygen. Another way is for intense stellar radiation to break apart water molecules into their constituent atoms, namely hydrogen and oxygen. The hydrogen, being very light, escapes into space but the oxygen, being much heavier, is retained to create an oxygen-rich atmosphere. This process is known as photodissociation (see Fig. 3.2). For this to work, there needs to be a lot of planetary water and a lot of stellar radiation, so such worlds are most likely to be both wetter and hotter than Earth.

I made this point in an article in 2013, calling these planets "spoof worlds" because they could have lots of oxygen in their atmospheres but still not possess any life [15]. Naturally, this concept was not original even then. In 2007, Antigona Segura, when she was a postdoc at Caltech, examined the possibility of whether spoof worlds could occur at relatively mild surface temperatures rather than the hot conditions that would be more favorable for photodissociation. Segura and her colleagues concluded that because oxygen was easily absorbed by surface compounds, on worlds with habitable temperatures it was unlikely that oxygen could build up fast enough to compensate for the considerably slower rate of spoof world production of oxygen on these cooler planets [16].

Since then, a number of studies have tried to estimate the potential range of planetary conditions under which spoof worlds could exist [17, 18]. A crucial concept in this work is whether the water can easily reach higher levels in the atmosphere where it can be broken apart by UV. On Earth, atmospheric temperature decreases with height in the lower part of the atmosphere, known as the troposphere. As air is almost completely transparent to sunlight but land and water are not, most of the heating occurs at the planetary surface where the sunlight is absorbed (Fig. 3.3). As the heated air rises from the

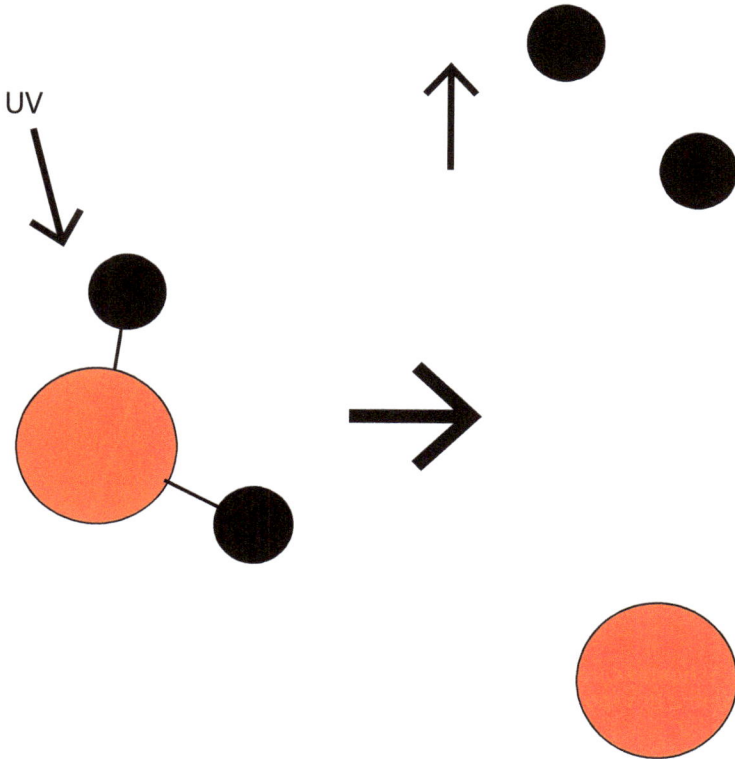

Fig. 3.2 The process of photodissociation. Incoming UV breaks apart a water molecule; the light hydrogen (small circles) escapes to space, while the heavier oxygen (large circles) remains

surface, its pressure decreases because gravity forces the air to collect preferentially in the lowest location, the surface, and thus the surface has the highest air pressure. As the pressure of a parcel of rising air decreases, it expands, for the same reason that it would compress if it were descending and were crushed by the higher pressure close to the surface. When the parcel expands, this requires some energy, and the only way to get that energy is from the molecules in the parcel themselves. Removing energy from them slows them down, and slower molecules means lower temperatures because that is what temperature is, a measure of the typical speed of molecular movement. So, putting this all together, at the top of the troposphere, temperature is considerably lower than at the surface.

Now it would be easy to assume that this decrease in temperature simply continues as the parcel rises and as the surrounding atmosphere disappears into outer space. But on Earth, above the troposphere the atmospheric temperature

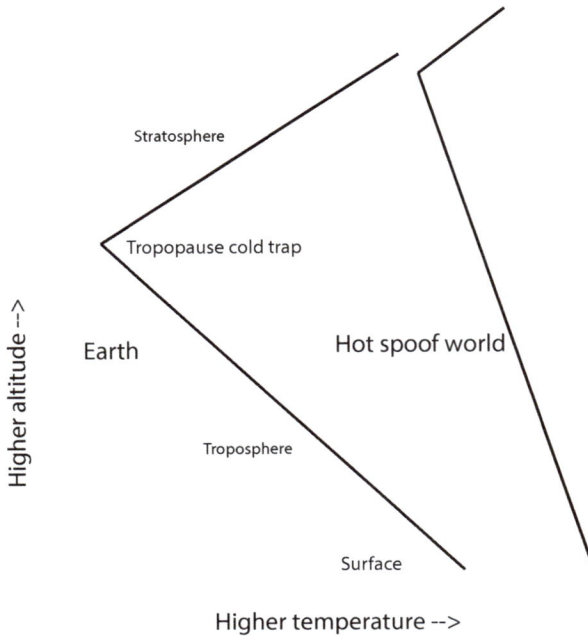

Fig. 3.3 Schematic vertical temperature profiles on Earth (left), indicating the cold trap, and on a hot spoof world (right), showing a much higher tropopause

starts to increase (Fig. 3.3). At these altitudes, in the stratosphere, ozone absorbs UV and heats up as a result. The minimum temperature at the top of the troposphere (the tropopause) creates what is known as the "cold trap", and it has important implications for the ability of the Earth to retain water. On Earth, the temperature at the cold trap is very far below the freezing point of water. By the time our parcel of air has risen to this point, almost all of its water has frozen out and dropped back towards the lower atmosphere. But on a much hotter world, there is a lot more surface heating and a lot more evaporation. The intense surface heating transports the water much higher in the atmosphere than on Earth, where it can be photodissociated by UV much more effectively. For instance, a planet with a mean surface temperature of about 70 °C (150 °F) would lose about 100 times as much water as Earth does, creating oxygen as it does so [19].

For a habitable spoof world to be created, this mechanism would need to work at moderate surface temperatures. Whether this can happen or not seems to depend on atmospheric composition. Work in 2013 and 2014 by Robin Wordsworth and Ray Pierrehumbert of the University of Chicago investigated the relationship between the cold trap and the fraction of the atmosphere that could condense at cold trap temperatures [17, 20].

Say a planet has an atmosphere comprised entirely of water vapor. In that case, there would be no cold trap because water vapor would percolate throughout the entire atmosphere, including the stratosphere, where it could be broken up into hydrogen and oxygen. Now add a gas that will not condense at Earth-like temperatures, like nitrogen. In this case, the nitrogen will just keep on cooling as it rises. In Fig. 3.3, this is represented by the lines on the left, showing the temperature change with height for Earth, a planet with a largely nitrogen atmosphere. As nitrogen is added to the atmosphere of a planet with a water vapor atmosphere, at some point low enough temperatures will occur in the troposphere for almost all of the water vapor to condense and thus for a cold trap to be formed. Small amounts of nitrogen do not seem to be enough to do this, though. Wordsworth and Pierrehumbert show that if the nitrogen concentration is less than a few percent of Earth's current concentration, the cold trap becomes ineffective and oxygen produced by photodissociation would start to build up the atmosphere. At the same time, if the carbon dioxide concentration were just right, the planet's greenhouse effect would give it temperate conditions at its surface, creating a spoof world at habitable temperatures.

But how likely is it? Not impossible, but it would require a number of specific conditions to occur, implying that these conditions might only be rarely satisfied, and that a temperate world with lots of oxygen in its atmosphere might actually be life-bearing instead [21]. Much more likely than temperate spoof worlds would be hot spoof worlds with big oceans. There would be no shortage of UV in the Vega system for the creation of these worlds, and while planets have not yet been detected in the system, they haven't been ruled out yet either. Tantalising evidence is provided by the Vega system's big dust content. Calculations suggest that it has been formed quite recently, probably by the break-up of an asteroid, creating what is known as a debris disk. There is a large gap in this disk between about 15 and 110 a.u. from Vega, perhaps indicating the presence of a planet or planets acting to sweep up dust in this region [22]. Since Vega is larger than our Sun, it could easily host one or more massive gas giants orbiting it, as our own smaller star the Sun is itself circled by two big Jovian-sized planets. It is also worth mentioning, though, that our Sun is not typical in this regard: only about 15–20% of G stars like our Sun are estimated to have one or more Jovians [23]. Instead, most G stars will have planets that are Neptune-sized or smaller. Our solar system is actually a bit of an odd man out in the known galaxy—more on this in Chap. 10.

*   *   *

The closest known planetary system in this sector comprises the two M stars of Gliese 752, located 5.9 parsecs away in Aquila, a bit to the southeast of Altair and only about a parsec distant from it. Circling Gliese 752A is a cold sub-Jovian discovered in 2018. Its orbit is reasonably eccentric so it may spend part of the time within the habitable zone [24], but the planet is too large to host terrestrial life. About 430 a.u. or so away from Gliese 752A is the smaller of the two stars in the system, Gliese 752B, which also goes by the name of VB 10 or van Biesbroeck's star. This is one of the least luminous M stars known and at the time of its discovery in 1944, it was the dimmest star ever found [25]. It is a particularly violent flare star of the UV Ceti type, named after another small M star located within the known galaxy that serves as the archetype of this kind of star (see Chap. 8).

Second on our itinerary in this sector is the planetary system of the K star HD 219134, located near the W in Cassiopeia and some 6.6 parsecs away from Earth. K stars are different in a number of ways from M stars. Apart from just being bigger, K stars are more stable and much less prone to flares. They are brighter than M stars, so planets in their habitable zones are considerably further away from their star than they are for M stars. This makes K star planets both less susceptible to atmosphere stripping and also less likely to be tidally locked. Overall, planets orbiting K stars have a number of built-in habitability advantages over similar planets orbiting M stars. Recent work has suggested, though, that K stars might experience a long period of high UV flux when they are young, possibly making them less habitable [26].

The star HD 219134 is both dimmer and much older than the Sun, with a venerable estimated age of 11 billion years [27]. It also has at least six planets (see Table 3.1), ranging in size from super-Earth to Jovian. We can get an idea of what some of these planets might look like even though they are too dim to be seen with current telescopes. As mentioned in Chap. 1, the temperature of a world affects the type of clouds it has, and cloud type can strongly influence the color and albedo of a planet. This idea is the basis of the Sudarsky classification for gas giant planets, which relates planetary temperature to the main cloud type that might be visible at the top of a gas giant atmosphere [28]. Stated simply, this classification suggests that cold Jovians like Jupiter may have relatively bright ammonia clouds, while slightly warmer worlds may have even brighter water clouds. Hot Jovians might have very dark sodium clouds and very hot gas giants might have iron or other metal clouds that are more reflective. At middling temperatures between about 350 and 800 K, Jovians would have few or no clouds and would have very low albedos as a result, due to the lack of suitable cloud particles that can condense at these temperatures.

Table 3.1 The planets of HD 219134, in order of distance from the star. Values from the NASA Exoplanet Archive [29]

Planet	Mass (Earth = 1)	Orbital period (days)	Distance from star (Earth distance from Sun = 1)	Equilibrium temperature (degrees K) (Earth = 255 K)	Type
HD 219134 b	4.74	3.1	0.039	925–1011	Very hot super-Earth
HD 219134 c	4.36	6.8	0.065	716–783	Hot super-Earth
HD 219134 f	>7.3	22.7	0.146	478–523	Hot sub-Neptune
HD 219134 d	>16.2	46.9	0.237	377–412	Hot sub-Jovian
HD 219134 g	>10.8	94.2	0.375	299–327	Hot sub-Jovian
HD 219134 h	>108	2247	3.11	104–113	Cold Jovian

How do Sudarsky's predictions compare with observations? We don't have many actual measurements of exoplanet colors or albedos to compare with Sudarsky's results. One planet, HD 189733 b, a very hot super-Jovian in this sector about 20 parsecs away, has had its color measured and found to be a deep, darkish blue, and may be cloudy rather than having a low albedo [30, 31]. Other recent results have suggested a more complicated relationship between temperature and albedo. Work published in 2022 by Raissa Estrela of CalTech and her co-authors uses analysis of observations from the Hubble Space Telescope of 62 exoplanets to show that there appears to be two populations of planets [32]. For one set, their atmospheres are clear across a wide range of temperatures. For another set, they are more cloudy or hazy at 500 and 2500 K, but less cloudy at about 1500 K. In 2024, Jonathan Brande of the University of Kansas and co-authors used the James Webb Space Telescope (JWST), launched in 2021, to examine the haziness of Neptune-sized sub-Jovians. They find that all have at least some cloud cover, but that there is a maximum of cloudiness between 500 and 700 K [33].

Based on these more recent results, we can speculate that the sub-Neptune HD 219134 f, with a temperature of about 1000 K, might lack clouds and so would be relatively dark. The sub-Jovian planet d is cooler, and so might start to display some reflectivity. The next planet out, planet g, might have some wispy white water clouds. Finally, the cold Jovian planet h may have ammonia clouds and might look like Jupiter or Saturn as a result. Note that these estimates are very rubbery as our understanding of exoplanet cloudiness and colors is rudimentary.

We'll briefly discuss the innermost, hottest planet, HD 219134 b, because it is the best observed, as it has both radial velocity and transit information that enable both its mass and radius to be determined accurately. It has more than four times the mass of the Earth and 1.5–1.6 times the radius, implying that it is a rocky planet rather than a gas giant [34, 35]. The planet is close enough to its star to likely be tidally locked, and its equilibrium temperature is higher than the melting points of several metals, including lead, zinc and magnesium. The probability that it has liquid water at its surface is just about zero. Surface gravity is intense, at about 1.8 times Earth's gravity or so. Even fit athletes would find it difficult to move freely on its surface, leaving aside the slight problem of tiptoeing around pools of molten metal. Its density is higher than Earth's, usually implying that like Earth it has a large core of metals such as iron. But there is a possible wrinkle here. Its density and distance from its star are also consistent with a composition largely dominated by calcium and aluminium minerals, a type of world previously predicted by numerical simulations of planet formation [36]. Writing in the paper proposing that HD 219134 b might be this type of planet [37], Caroline Dorn of the University of Zurich waits until the very last line to make the following sly point: "On a popular science note, these worlds are rich in sapphires and rubies", since these gemstones are compounds of aluminium.

In journalism, this is known as "burying the lead". Accordingly, we will now call HD 219134 b "Sapphire World".

Sapphire World likely has an atmosphere, although it is not known how thick it is. We do suspect, however, that its atmosphere is unlikely to be composed of its original hydrogen and helium. A hydrogen atmosphere so close to its star will be strongly heated and will rapidly escape unless it is very thick to start with. The atmosphere of Sapphire World as calculated from Dorn's model is too thin to be both composed of hydrogen and to still be there after 11 billion years. Instead, Sapphire World almost certainly has a "secondary" atmosphere generated by massive volcanic eruptions over millions of years, comprising gases such as carbon dioxide and water vapor. If its internal composition is dominated by calcium and aluminium instead of silicon and iron, it would not have a strong magnetic field, because it would need lots of iron for that. A weak magnetic field makes its atmosphere vulnerable to stripping. Also, since the planet is old, rates of volcanic eruption would also be lower than for a younger planet, meaning less replenishment of its atmosphere. On the other hand, the planet is much larger than Earth so it might have had more volcanoes to begin with and thus a bigger atmosphere. On the third hand, the side facing the star would be so hot that an atmosphere of sorts would be created just from the evaporation of bits of the molten surface. Leaving aside the pleasing prospect of a king's ransom in precious gems lying on the ground just there for the taking, Sapphire World would be a very nasty place. If it has an atmosphere as thick as Earth's, it would have equally nasty, high-energy weather.

HD 219134 b *How similar to Earth: Not at all. A very hot super-Earth.*
Most probable planet: Tidally locked, very hot super-Earth with partially molten surface on the starward side.
Best case scenario: Much the same.
Worst case scenario: Ditto.
Cultural connection: Sapphire World.

*   *   *

Altair is an A star that rotates rapidly, about once every 9 h, and like Vega is egg-shaped as a result. We know this for sure because Altair is one of few stars that has been directly imaged at sufficient resolution to resolve its eggy structure [38]. So far, no planets have been found orbiting it, and with Altair's age of about 100 million years, any planets in its habitable zone may not have developed sophisticated lifeforms yet. Like Vega, Altair is a nearby bright star, so science fiction writers have gone to town in describing its planets [39]. Only a couple of examples will be mentioned here. One of the first was the classic film *Forbidden Planet* (1956), featuring the debut of Robby the Robot, along with other special effects that are still impressive even by today's standards [40]. The film is set in the rather desolate-looking landscape of the habitable planet Altair IV, a terrain reminiscent of the southwestern United States, like that of so many other fictional film planets. In Hal Clement's *Close to Critical* (1958) [41], the planet Tenebra in the Altair system is a hot super-Earth with a surface temperature of about 374 °C and an atmosphere containing both water vapor and sulphur. The suspiciously precise value of 374 degrees is deliberately chosen so that it is close to what is known as the "critical point" of water, occurring at this temperature and at a surface pressure of about 200 times that of Earth. At higher temperatures and pressures than this, H_2O is neither a gas nor a liquid but is instead a supercritical fluid. In a liquid, molecules roll around each other, as they are not moving fast enough to escape completely from the forces that attract them to each other. In a gas, molecules are fast enough to escape, so they speed off every which way. In a supercritical fluid, the temperature is so high that the molecules are moving too fast to be turned into a liquid, no matter how much they are compressed into each other by high atmospheric pressure. The result is that a supercritical fluid has a higher density than a gas but usually a lower density than a liquid. On Tenebra, at night a slight cooling of its atmosphere below the critical point leads to rapid rainout of water vapour in enormous, meters-wide raindrops. Supercritical fluids are found in the atmospheres of several solar system planets, including Jupiter, Saturn and Venus. On Venus, the presence of a supercritical carbon dioxide fluid at the surface may well be affecting its meteorology. Usually, an atmosphere with a very strong decrease of temperature with height will be vertically unstable, meaning that hot parcels of gas will rocket up from the surface into the much cooler higher levels of the atmosphere, one of the mechanisms that causes thunderstorms on Earth. But on Venus, despite some measurements showing a strong vertical temperature decrease, this instability is not occurring. The answer to this paradox could be that the lower atmosphere on Venus is a supercritical fluid that is considerably denser than a gas. If the atmosphere near the surface is supercritical but cools and transforms to a gas at higher levels, the much denser supercritical fluid will hug the surface and not become unstable [42]. A similar phenomenon could occur on Tenebra.

Finally, in *The Hitchhiker's Guide to the Galaxy* (1979), Douglas Adams informs us that the Altairian dollar is a widely accepted currency throughout the galaxy and thus very helpful to any backpackers wandering through the Milky Way [43]. One suspects that a tangible medium of exchange might be more generally useful—like sapphires, maybe.

There are plenty of other well-known systems in this sector within 10 parsecs of Earth. The 70 Ophiuchi system consists of two habitable-friendly 2 billion year old K stars orbiting between 11 and 35 a.u. from each other [44]. No planets have yet been found, however. A system where there are almost certainly no habitable planets is that of Chi Draconis [45]. Its two decent-sized, approximately 5 billion year old stars circle each other on an eccentric orbit with a typical separation of about 1 a.u. [46], making it just about gravitationally impossible for any planet to exist within either star's habitable zone. Habitable planets circling around both stars (instead of just one of them) are probably unlikely as well, as for them to be habitable they would likely have to be too close to the stars for their orbits to be stable [47] (sorry, *Babylon 5*—the cold habitable planet Minbar will have to be located somewhere else [48]). In the same constellation, Sigma Draconis is another star well known to science fiction fans [49]. In the strange *Star Trek* episode *Spock's Brain*, the crew of the starship Enterprise find themselves on the ice planet Sigma Draconis VI, in search of an alien who has stolen Spock's brain (the B movie plot gives you an idea of the quality of this episode) [50]. While this 3 billion year old K star is perfectly suitable for habitable planets, no planets have yet been found [51].

The 61 Cygni system is famous for a number of reasons. In 1838, Friedrich Bessel measured its parallax, and thus this was the first star whose distance was reliably determined. Its two mature K star components are separated by between 44 and 124 a.u., far enough apart so that they could easily both host a number of planets each. Indeed, recent measurements suggest a super-Jovian planet orbiting 61 Cygni B at a distance of between 10 and 20 a.u. [52] Still, despite this system being a high priority target for planetary searches, no real planets have been confirmed, but the system hosts a bunch of fictional ones [53]. Hal Clement's very inventive novel *Mission of Gravity* (1954) describes the cold super-Jovian world Mesklin, a planet of 61 Cygni A that rotates so fast that the surface gravity at the equator is "only" three times that of Earth, while at the pole it is several hundred times [54]. On this world, the apparent gravity is less at the equator because centrifugal force from the rapid spinning of the planet throws objects outwards and thus partly counteracts the downward force of gravitational attraction. In the novel *Chasm City* (2001) by Alastair Reynolds, this system hosts the war-torn terrestrial jungle planet Sky's Edge [55]. The lush vegetation of the planet, while superficially familiar to Earth's, is inedible as it induces anaphylaxis in anyone who tries to eat it. In C.J. Cherryh's Alliance-Union universe, which includes the Hugo award-winning novel *Downbelow Station* (1981), the 61 Cygni system contains a station on the Great Circle trade route, Bryant's Star Station [56]. There are other stars with such stations in this sector, including the M stars Kruger 60 and Ross 248. In Cherryh's version of the future, these stations are slowly constructed during the period when spaceships could only travel slower than light. Communications with Earth are tediously slow, and the stations effectively become independent outposts. But then in 2248, faster-than-light travel is developed, Earth tries to re-establish control over the stations, and rebellion ensues.

Also in this sector is the M star Lalande 46650, the home system of the planet Cyteen, headquarters of the Union, the arch enemy of the Earth-based Alliance in Cherryh's fictional universe. Cyteen is not an hospitable world, requiring non-stop artificial maintenance of some parts of its surface so that these regions can remain habitable. In reality, its star is moderately deficient in heavy elements like iron, which tends to imply a planetary system comprised of smaller worlds, if they exist. Further out in this sector and a little north of the constellation Pegasus is the young, small flare star EV Lacertae. This star is noteworthy because in 2008, it released a flare that increased its visible brightness by a factor of almost a hundred [57]. This would have had potentially disastrous consequences for the atmosphere and life of any habitable zone planets in this system [58].

Finally, there are two other stars worth talking about. The first is the M star or brown dwarf LSR J1835+3259, which based on some recent observations might have auroras [59]. The other stellar system is that of GJ 1230, consisting of three M stars, with two orbiting each other in only a few days, and the third one considerably further away [60]. Not much is known about these systems, so why have I even mentioned them? These two nearby stellar systems lie on a kind of direct interstellar route between Vega and Altair, in other words between our pair of separated lovebirds who only meet once a year on the festival day of Qixi. These stars are the real magpie bridge across the starry Milky Way.

Further Afield

Just outside our 10 parsec limit in this sector is 51 Pegasi, a 5 billion year old G star that has recently left the main sequence and will continue to become brighter on its journey toward red giantdom [61]. Its only known planet used to be called 51 Pegasi b but now goes by the name of Dimidium. Dimidium is Latin for "half" and refers to the mass of the planet being half that of Jupiter. It is famous because in 1995 it was the first extrasolar planet discovered orbiting a main-sequence star [62]. For this discovery, Michel Mayor and Didier Queloz of the University of Geneva shared the 2019 Nobel Prize for Physics. Some planets had previously been discovered orbiting a neutron star, but since neutron stars emit a torrent of lethal radiation, those planets are generally (but maybe not entirely) an unhealthy environment for life [63]. Dimidium is a very hot Jovian and its discovery was a huge surprise because at that time, according to then well-accepted theories of planetary formation and evolution, large planets were not supposed to orbit so close to their primary star. The discovery of Dimidium motivated the gradual realization that while planets may form at a certain distance from their star, they do not always stay there and can migrate inwards and sometimes outwards. Dimidium may have formed further away from 51 Pegasi and then moved inwards early in the development of the system, on the way ejecting smaller planets that are now wandering the galaxy as Steppenwolves [64]. While water vapor has been detected in Dimidium's atmosphere, because the planet has day side temperatures well over 1500 K, it also likely has metallic clouds high in its atmosphere—perhaps even sapphire clouds—along with a wild atmospheric circulation with strong equatorial winds [65–67].

Big hot Jovians like Dimidium are easy to discover with modern techniques. But the history of the detection of the hot super-Jovian planet Tadmor, in the Gamma Cephei system in this sector about 14 parsecs away from Earth, is a lesson in how scientific progress sometimes takes two steps forward and one step back. Gamma Cephei A, a K subgiant about 11 times brighter than our Sun, was once an F or A main sequence star but is now evolving towards giant status. It is therefore termed "a retired A star" [68]. In 1988, Bruce Campbell, Gordon Walker and collaborators from the University of Victoria, British Columbia, tentatively identified an unexplained radial velocity signal in data from this star that indicated a planet of at least 1.7 Jupiter masses [69]. As this would have been the first extrasolar planet to be discovered, it would have been huge news. Unfortunately, in 1992 the authors retracted this claim because they became concerned that the measured radial velocity variations could actually be variations within the star itself rather than from the planet. Nevertheless, further measurements were made, and in 2002 this planet, now known as Tadmor, was confirmed with a minimum mass of 1.85 Jupiter masses. The original measurements made by the Canadian group were not bad after all [70].

Tadmor is the ancient name for the Greek city of Palmyra, former capital of the short-lived third century Palmyrene empire, whose extensive ruins can still be seen in Syria today [71]. The planet orbits about 2 a.u. from Gamma Cephei A. Studies of Tadmor have focused on the possible gravitational instability of its stellar system. Gravitational instability is caused by the gravitational attraction of planets or stars causing small variations in the orbit of a planet, variations that grow over time until they are large enough to change its orbit. Known as perturbations, these wobbles can become big enough for a planet to be completely ejected from its stellar system. In the Gamma Cephei system, this is not an idle concern, as there is an M star orbiting Gamma Cephei A that approaches to about 10 a.u. from the bigger star, uncomfortably close to Tadmor. Nevertheless, calculations suggest that the Gamma Cephei system is gravitationally stable if the M star orbits in the same plane as the orbit of Tadmor [72].

Calculations of whether a system is gravitationally stable or not have also been used to get a better handle on the likely masses and orbits of some other stellar systems. For instance, the young, 40 million year old HR 8799 system is also in this sector but 40 parsecs away, in very nearly the same direction as 51 Pegasi. The HR 8799 system contains planets still youthful enough to be glowing with leftover heat from their formation, making them detectable by infrared telescopes [73]. They are big planets but because their orbits take decades to complete, it is difficult to assign masses to them because radial velocity measurements cannot be easily made for planets with such long orbits. These measurements rely on astronomers observing the to-and-fro gravitational tug of the planet as it circles around its star, and if the to-and-fro takes 50 years, well, you see the problem. Enter stability analysis: the laws of gravitation dictate that only certain combinations of orbits and planetary masses in this system are gravitationally stable. Taking this into account, calculations indicate that the planets of the HR 8799 system are all super-Jovians with masses in the range of five to seven Jupiter masses [74]. Also, interestingly, the system is much more stable if orbital periods are integer multiples of each other. In the HR 8799 system, the planets in order of increasing distance from the star are planet e, planet d, planet c and planet b. Stable orbits are achieved if the period of planet d is twice that of e, if the period of c is twice that of d, and if the period of b is twice that of c, a phenomenon known as orbital resonance. There are other resonances that are stable in this system as well, but the point here is that there is no automatic guarantee of gravitational stability in a stellar system, particularly during the early part of its evolution. Planets are ejected from stellar systems frequently.

Another claim to fame for this sector is the area to the northeast of Deneb, known as the Kepler Field of View. This is the region of the sky that was surveyed by the Kepler space telescope, now defunct, whose job was to find transiting planets by simultaneously monitoring the brightness of a large number of stars. Thousands of planets were detected this way, and thousands more possible planets found by Kepler are yet to be confirmed by further analysis of the data [75]. As a result, in this sector there are a very large number of exoplanets detected by Kepler. None of them are closer than 10 parsecs, however. The closest Kepler planets detected so far in this region are in the Kepler 444 system, a triple star system some 36 parsecs away comprising an ancient 11 billion year old K star and a pair of M stars [76, 77]. The K star has five known rocky and sub-rocky planets, all far too hot to be habitable. This pattern of a stellar system consisting of several small rocky and sub-rocky hot planets is repeated for a number of the Kepler-detected systems. This is further evidence that smaller, older metal-poor stars form a number of small planets in close orbits rather than giant planets. Large worlds instead tend to form in systems that are younger and more metal-rich, like our own solar system [78]. In the case of Kepler 444, formation of large planets orbiting the main K star is not aided by the presence of the M stars, as they may approach the K star to within 5 a.u.. This would wreck the formation of any planets orbiting at about 1 a.u. or so from the K star [79]. Later work has revised the closest approach to about 20 a.u., though.

Another possible way to wreck a stellar system is when a star evolves into a red giant. The resulting increase in the brightness of a solar-type star, by more than 1000 times, would vaporize some inner planets and drastically change the environments of all the others [80]. In our solar system, when our Sun becomes a red giant, even Jupiter will not be immune, with our current cold Jovian planet becoming a hot Jovian [81]. When the Sun reaches its maximum brightness as a red giant, the habitable zone of the solar system will move out to the Kuiper belt. Still, big planets can survive this phase without losing much of their atmospheres [82]. We have direct evidence for this in the form of the hot super-Jovian planet Fortitudo ("fortitude"), orbiting the G giant Xi Aquilae in this sector some 56 parsecs away [83]. This world certainly is displaying its fortitude as is still hanging in there, despite now being blasted with more than 100 times the incoming stellar radiation that Earth experiences.

Finally, way, way out at about 800 parsecs is Deneb, our destination for heavenly magpies. Deneb used to be a main sequence O star but is now an A type supergiant, and supergiant stars live fast and die young [84]. In only a few million years, there will be a giant bang and Deneb will explode, leaving behind a neutron star, a black hole—or maybe nothing at all.

Vega Sector Summary

Qixi, Vega, spoof worlds, Sapphire World, Cyteen, the real magpie bridge, Dimidium and its sapphire clouds, Keplers galore, and doomed Deneb: there is more to talk about in the Vega sector than can easily be included in our brief survey of the known galaxy. There is more that can be said about Vega itself. For instance, among many indigenous peoples of northern Asia and North America, there is a geographically widespread belief that the world is supported by a giant pillar that runs up through the pole star, the so-called "world pillar". This is a reasonable assumption for peoples who spent a lot of time at high northern latitudes, where the pole star is very high in the sky and the stars appear to rotate around it. Other cultures world-wide have a similar idea, the so-called axis mundi, a special location connected to the heavens by a great tree or pillar. Scholars are divided regarding how old this idea is, but like many ancient myths it is perfectly possible that it originated in paleolithic times and thus is thousands or tens of thousands of years old [85]. Since the myth of a world pillar running through the pole star is told among both Siberian and Native American peoples, and since these populations are closely related genetically, it seems unlikely that this concept originated completely independently in these two related cultures. More likely, we can speculate that the tale was transported from Siberia to North America when the Native Americans moved there. Thus the myth of the polar world pillar may predate the arrival of First Nations peoples in North America more than 10,000 years ago [86]. The pole star is currently Polaris, but the location of the celestial north pole slowly moves as the Earth wobbles on its axis over periods of thousands of years. In 12,000 B.C.E., the stellar north pole was in the constellation Lyra [87]. So it is quite possible that when the idea of the polar world pillar originated, the indigenous people of Siberia and their descendants in North America did not mean that the pillar ran through Polaris: they meant that it ran through Vega.

References

1. Brown J, Brown J (2006) China, Japan, Korea: culture and customs. BookSurge, North Charleston, SC, p 72
2. Davis AR (ed) (1970) The Penguin book of Chinese verse. Penguin, Baltimore, MD
3. Jones A (2021) China is working on a relay satellite to support lunar polar missions. spacenews.com. https://spacenews.com/china-is-working-on-a-relay-satellite-to-support-lunar-polar-missions/. Accessed 8 Feb 2024
4. Knobel EB (1895) Al Achsasi Al Mouakket, on a catalogue of stars in the Calendarium of Mohammad Al Achsasi Al Mouakket. Mon Not Roy Astronom Soc 55(8):429–438. https://doi.org/10.1093/mnras/55.8.429
5. Yoon J et al (2010) A new view of Vega's composition, mass, and age. Astrophys J 708(1):71–79
6. Asimov I (1951) Foundation. Gnome, New York
7. Blish J (1970) Cities in flight (omnibus ed). Avon, New York
8. Heinlein RA (1958) Have space suit—will travel. Scribner's, New York
9. Vance J (1964) Star king (Demon princes #1). Berkley, New York
10. Sagan C (1985) Contact. Simon and Schuster, New York
11. Cnossen I, Sanz-Forcada J, Favata F et al (2007) Habitat of early life: Solar X-ray and UV radiation at Earth's surface 4–3.5 billion years ago. J Geophys Res Planet 112:E2
12. Guilloux C-A, Lamoureux C, Beauruelle C, Héry-Arnaud G (2021) Porphyromonas: A neglected potential key genus in human microbiomes. Anaerobe 68:102230. https://doi.org/10.1016/j.anaerobe.2020.102230
13. O'Conner C (2014) Cell function depends on the continual uptake and conversion of energy. In: Essentials of cell biology. https://www.nature.com/scitable/topicpage/cell-energy-and-cell-functions-14024533/. Accessed 9 Feb 2024
14. Cardona T (2018) Early Archean origin of heterodimeric photosystem I. Heliyon 4(3):e00548. https://doi.org/10.1016/j.heliyon.2018.e00548
15. Walsh K (2013) Spoof worlds. Analog science fiction and fact, March 2013
16. Segura A, Meadows VS, Kasting JF et al (2007) Abiotic formation of O_2 and O_3 in high-CO_2 terrestrial atmospheres. Astronom Astrophys 472(2):665–679
17. Wordsworth R, Pierrehumbert R (2014) Abiotic oxygen-dominated atmospheres on terrestrial habitable zone planets. Astrophys J Lett 785(2):L20

18. Wordsworth R, Kreidberg L (2022) Atmospheres of rocky exoplanets. Ann Rev Astronom Astrophys 60:159–201

19. Kasting J (2010) How to build a habitable planet. Princeton University Press, Princeton, NJ

20. Wordsworth RD, Pierrehumbert RT (2013) Water loss from terrestrial planets with CO_2-rich atmospheres. Astrophys J 778(2):154

21. Ramirez RM (2018) A more comprehensive habitable zone for finding life on other planets. Geosciences 8(8):280

22. Su KY, Rieke GH, Malhotra R et al (2013) Asteroid belts in debris disk twins: Vega and Fomalhaut. Astrophys J 763(2):118. https://doi.org/10.1088/0004-637X/763/2/118

23. Cumming A, Butler RP, Marcy GW et al (2008) The Keck planet search: detectability and the minimum mass and orbital period distribution of extrasolar planets. Publ Astronom Soc Pac 120(867):531–554. https://doi.org/10.1086/588487

24. Kaminski A, Trifonov T, Caballero JA et al (2018) The CARMENES search for exoplanets around M dwarfs—a Neptune-mass planet traversing the habitable zone around HD 180617. Astronom Astrophys 618:A115

25. van Biesbroeck G (1944) The star of lowest known luminosity. Astronom J 51:61–62. https://doi.org/10.1086/105801

26. Richey-Yowell T, Shkolnik EL, Loyd RP et al (2022) HAZMAT. VIII. A spectroscopic analysis of the ultraviolet evolution of K stars: additional evidence for K dwarf rotational stalling in the first gigayear. Astrophys J 929(2):169

27. Seager S, Knapp M, Demory BO et al (2021) HD 219134 revisited: planet d transit upper limit and planet f transit nondetection with ASTERIA and TESS. Astronom J 161(3):117

28. Sudarsky D, Burrows A, Pinto P (2000) Albedo and reflection spectra of extrasolar giant planets. Astrophys J 538(2):885–903. https://doi.org/10.1086/309160

29. NASA (2023) HD 219134 overview. In: NASA Exoplanet Archive. https://exoplanetarchive.ipac.caltech.edu/cgi-bin/DisplayOverview/nph-DisplayOverview?objname=HD+219134&type=PLANET_HOST. Accessed 10 Oct 2023

30. Vallenari A et al (2023) Gaia Data Release 3. Summary of the content and survey properties. Astronom Astrophys 674:A1. https://doi.org/10.1051/0004-6361/202243940

31. Evans TM, Pont F et al (2013) The deep blue color of HD 189733 b: albedo measurements with Hubble Space Telescope/Space Telescope Imaging Spectrograph at visible wavelengths. Astrophys J Lett 772(2):L16. https://doi.org/10.1088/2041-8205/772/2/L16

32. Estrela R, Swain MR, Roudier GM (2022) A temperature trend for clouds and hazes in exoplanet atmospheres. Astrophys J Lett 941(1):L5

33. Brande J, Crossfield IJ, Kreidberg L et al (2024) Clouds and clarity: revisiting atmospheric feature trends in Neptune-size exoplanets. Astrophys J Lett 961(1):L23

34. Motalebi F et al (2015) The HARPS-N rocky planet search. I. HD 219134 b: A transiting rocky planet in a multi-planet system at 6.5 pc from the Sun. Astronom Astrophys 584:A72. https://doi.org/10.1051/0004-6361/201526822
35. Gillon M et al (2017) Two massive rocky planets transiting a K-dwarf 6.5 parsecs away. Nature Astronom 1(3):0056. https://doi.org/10.1038/s41550-017-0056
36. Thiabaud A, Marboeuf U, Alibert Y et al (2014) From stellar nebula to planets: The refractory components. Astronom Astrophys 562:A27
37. Dorn C, Harrison JH, Bonsor A, Hands TO (2019) A new class of Super-Earths formed from high-temperature condensates: HD219134 b, 55 Cnc e, WASP-47 e. Mon Not Roy Astronom Soc 484(1):712–727
38. Monnier JD, Zhao M, Pedretti E et al (2007) Imaging the surface of Altair. Science 317(5836):342–345
39. Wikipedia (2023) Altair. Accessed 11 Oct 2023
40. Forbidden Planet (1956) imbd.com. https://www.imdb.com/title/tt0049223/?ref_=nv_sr_srsg_0_tt_8_nm_0_q_forbidden%2520planet. Accessed 11 Oct 2023
41. Clement H (1958) Close to critical. Ballantine, New York City
42. Lebonnois S, Schubert G (2017) The deep atmosphere of Venus and the possible role of density-driven separation of CO_2 and N_2. Nat Geosci 10(7):473–477. https://doi.org/10.1038/NGEO2971
43. Adams D (1979) The hitchhiker's guide to the galaxy. Pan, London
44. Wikipedia (2023) 70 Ophiuchi. Accessed 12 Oct 2023
45. Wikipedia (2023) Chi Draconis. Accessed 12 Oct 2023
46. Pourbaix D (2000) Resolved double-lined spectroscopic binaries: A neglected source of hypothesis-free parallaxes and stellar masses. Astronom Astrophys Supp Ser 145(2):215–222
47. Musielak ZE, Cuntz M, Marshall EA, Stuit TD (2005) Stability of planetary orbits in binary systems. Astronom Astrophys 434(1):355–364
48. Babylon 5 (1993) imdb.com. https://www.imdb.com/title/tt0105946/. Accessed 14 Oct 2023
49. Wikipedia (2023) Sigma Draconis. Accessed 18 Oct 2023
50. Star Trek (1966) Spock's brain. Sept 20 1968. imdb.com. https://www.imdb.com/title/tt0708449/?ref_=ttep_ep1. Accessed 18 Oct 2023
51. Ramírez I et al (2012) Lithium abundances in nearby FGK dwarf and subgiant stars: internal destruction, galactic chemical evolution, and exoplanets. Astrophys J 756(1):46. https://doi.org/10.1088/0004-637X/756/1/46
52. Kervella P, Arenou F, Thévenin F (2022) Stellar and substellar companions from Gaia EDR3-proper-motion anomaly and resolved common proper-motion pairs. Astronom Astrophys 657:A7
53. Wikipedia (2023) 61 Cygni. Accessed 17 Oct 2023
54. Clement H (1979) Mission of gravity. Doubleday, New York
55. Reynolds A (2001) Chasm city. Gollancz, London
56. Cherryh CJ (1981) Downbelow station. DAW, New York

57. Paudel RR, Barclay T, Schlieder JE et al (2021) Simultaneous multiwavelength flare observations of EV Lacertae. Astrophys J 922(1):31
58. Osten RA, Godet O, Drake S et al (2010) The mouse that roared: a superflare from the dMe flare star EV Lac detected by Swift and Konus-Wind. Astrophys J 721(1):785
59. Hallinan G, Littlefair SP, Cotter G et al (2015) Magnetospherically driven optical and radio aurorae at the end of the stellar main sequence. Nature 523(7562):568–571
60. Gizis JE, Reid NI (1996) GJ 1230: a nearby triple system. Astronom J 111:365
61. Wikipedia (2023) 51 Pegasi b. Accessed 19 Oct 2023
62. Mayor M, Queloz D (1995) A Jupiter-mass companion to a solar-type star. Nature 378(6555):355–359. https://doi.org/10.1038/378355a0
63. Patruno A, Kama M (2017) Neutron star planets: atmospheric processes and irradiation. Astronom Astrophys 608:A147
64. Lin DN, Bodenheimer P, Richardson DC (1996) Orbital migration of the planetary companion of 51 Pegasi to its present location. Nature 380(6575):606–607
65. Brogi M, Snellen IAG, de Kok RJ et al (2013) Detection of molecular absorption in the dayside of exoplanet 51 Pegasi b? Astrophys J 767(1):27
66. Parmentier V, Fortney JJ, Showman AP et al (2016) Transitions in the cloud composition of hot Jupiters. Astrophys J 828(1):22
67. Kataria T, Sing DK, Lewis NK et al (2016) The atmospheric circulation of a nine-hot-Jupiter sample: probing circulation and chemistry over a wide phase space. Astrophys J 821(1):9
68. Johnson JA, Huber D, Boyajian T et al (2014) The physical parameters of the retired A star HD 185351. Astrophys J 794(1):15
69. Campbell B et al (1988) A search for substellar companions to solar-type stars. Astrophys J 331:902–921. https://doi.org/10.1086/16660
70. Hatzes AP et al (2003) A planetary companion to Gamma Cephei A. Astrophys J 599(2):1383–1394. https://doi.org/10.1086/379281
71. O'Connor MP (1988) The etymologies of Tadmor and Palmyra. In: Arbeitman YL (ed) A linguistic happening in memory of Ben Schwartz: studies in Anatolian, Italic, and other Indo-European languages. Bibliothèque des cahiers de l'institut de linguistique de Louvain (BCILL) 42. Peeters, Leuven
72. Torres G (2007) The planet host star γ Cephei: physical properties, the binary orbit, and the mass of the substellar companion. Astrophys J 654(2):1095–1109. https://doi.org/10.1086/509715
73. Marois C, Macintosh B, Barman T (2008) Direct imaging of multiple planets orbiting the star HR 8799. Science 322(5906):1348–1352. https://doi.org/10.1126/science.1166585
74. Wang JJ, Graham JR, Dawson R et al (2018) Dynamical constraints on the HR 8799 planets with GPI. Astronom J 156(5):192
75. Borucki WJ (2016) KEPLER Mission: development and overview. Rep Prog Phys 79(3):036901

76. Campante TL, Barclay T, Swift JJ (2015) An ancient extrasolar system with five sub-Earth-size planets. Astrophys J 799(2):170
77. Buldgen G et al (2019) Revisiting Kepler-444. I. Seismic modeling and inversions of stellar structure. Astronom Astrophys 630:A126. https://doi.org/10.1051/0004-6361/20193612
78. Buchhave LA, Latham DW, Johansen A et al (2012) An abundance of small exoplanets around stars with a wide range of metallicities. Nature 486(7403):375–377
79. Dupuy TJ et al (2016) Orbital architectures of planet-hosting binaries. I. Forming five small planets in the truncated disk of Kepler-444A. Astrophys J 817(1):80. https://doi.org/10.3847/0004-637X/817/1/80
80. Schröder K-P, Connon Smith R (2008) Distant future of the Sun and Earth revisited. Mon Not Roy Astronom Soc 386(1):155–163. https://doi.org/10.1111/j.1365-2966.2008.13022.x
81. Spiegel DS, Madhusudhan N (2012) Jupiter will become a hot Jupiter: consequences of post-main-sequence stellar evolution on gas giant planets. Astrophys J 756(2):132
82. Passy JC, Mac Low MM, De Marco O (2012) On the survival of brown dwarfs and planets engulfed by their giant host star. Astrophys J Lett 759(2):L30
83. Sato B et al (2008) Planetary companions around three intermediate-mass G and K giants: 18 Delphini, ξ Aquilae and HD 81688. Publ Astronom Soc Japan 60(3):539–550. https://doi.org/10.1093/pasj/60.3.539
84. Wikipedia (2023) Deneb. Accessed 20 Oct 2023
85. Pentikäinen J (ed) (1996) Shamanism and northern ecology. De Gruyter, Oldenbourg
86. Goebel T, Waters MR, O'Rourke DH (2008) The Late Pleistocene dispersal of modern humans in the Americas. Science 319(5869):1497–1502. https://doi.org/10.1126/science.1153569
87. Dyches P (2021) What is the North Star and how do you find it? NASA. https://science.nasa.gov/solar-system/skywatching/what-is-the-north-star-and-how-do-you-find-it/. Accessed 24 Oct 2023

4

The Guardian of the Bear

The Arcturus Sector: 12–18 h, 0 to +90°

The next star on our tour is one that can't even be seen with the naked eye. Barnard's Star (Fig. 4.1) is the second-nearest stellar system to Earth. It was named after American astronomer E.E. Barnard, who in 1916 found that it was moving very rapidly across the sky relative to more distant stars, giving it a high "proper motion". This is usually a sure sign that a star is nearby [1]. Naturally, "very rapidly" in this context means that it takes more than 150 years to shift its position in the sky by only the apparent distance of the width of the Moon, but that is positively zooming along by stellar standards. Barnard's Star has the highest known proper motion.

Barnard's Star was not known by the ancients, but twentieth century science fiction writers have written a lot about it. It is the site of Alpha Station, part of the series of space stations constructed in the early days of interstellar exploration, as described in C.J. Cherryh's Alliance-Union series. It is the home star of Barnard's World, the setting of Dan Simmons's very inventive novel *Hyperion* (1989), where the star hangs in the sky like an enormous red orb [2]. This Hugo award-winning novel follows the same format as *The Canterbury Tales* by Chaucer and comprises a series of stories narrated by members of a band of pilgrims on a journey to a dangerous alien artifact. This system is also the location of *Rocheworld* (1990), where two planets are so close together that they share an atmosphere [3]. This is a brilliantly original concept and technically possible, but as others have suggested it is rather unlikely, as there are many factors that might cause this situation to become unstable.

In 2018, a super-Earth planet was announced orbiting Barnard's Star, discovered using the radial velocity method [4]. This was very exciting news, but

© The Author(s), under exclusive license to Springer Nature Switzerland AG 2024
K. J. E. Walsh, *Planets of the Known Galaxy*, Science and Fiction,
https://doi.org/10.1007/978-3-031-68218-6_4

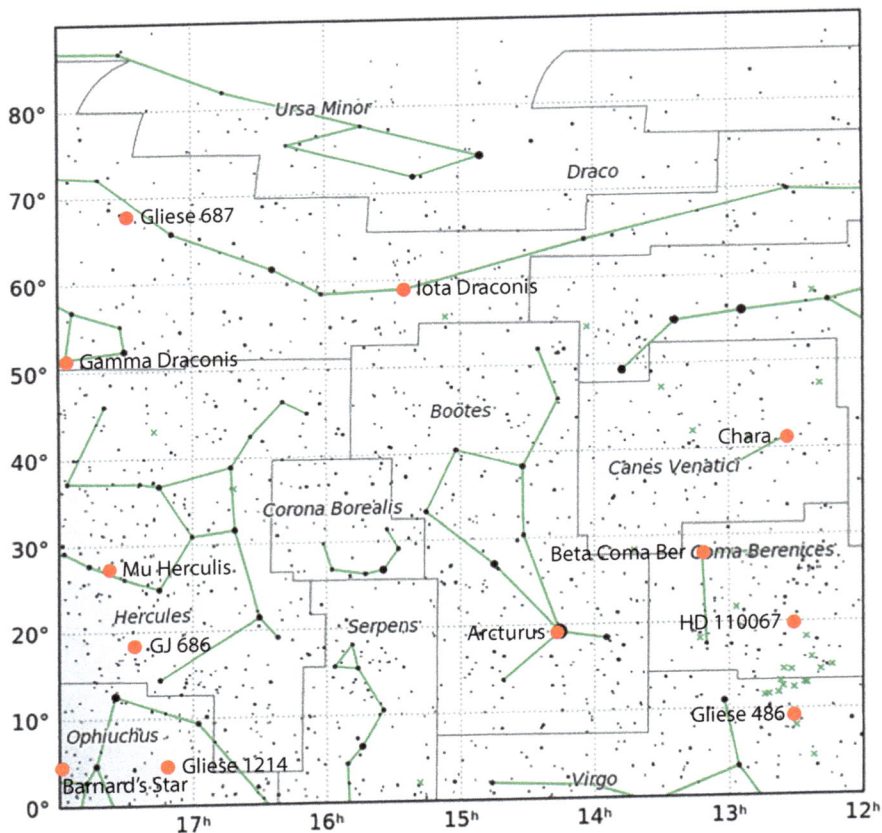

Fig. 4.1 Map of the Arcturus sector. Blue shading indicates the "Milky Way", the most star-packed part of the galaxy, while the blue (darker) line is the galactic equator, an imaginary circle running through the Milky Way that slices the galaxy in half. The orange (lighter) line is the path of the Sun's location through the seasons (indicating the "zodiac" constellations). Larger circles indicate brighter stars, while locations mentioned in this chapter are indicated by red circles. Image produced using StarCharter (https://github.com/dcf21/star-charter, accessed on the 2nd of July 2024. © 2020, Dominic Ford) software under the GNU general public license

less exciting was a report in 2021 suggesting that this signal was a false positive [5]. When variations are detected in a star's frequency, one possible cause is something associated with the rotation of the star itself, like a starspot. This can mimic the radial velocity variations caused by a planet and is now proposed as a cause of the apparent detection of this planet. Still, we'll just press on and assume, for the sake of argument, that the planet is real. This is not such a bad assumption because worlds like this are almost certain to exist around other stars in the known galaxy or elsewhere. Since this hypothetical world is a satellite of Barnard's Star, I'll call it "Columbia".

Now Columbia is very close to the snow line, the distance from the star beyond which it is cold enough for water vapor, carbon dioxide and other so-called "ices" to remain solid and accumulate rather than evaporate into space [4]. The significance of the snow line is that the accumulation of ices makes it much easier for planets to grow during their formation. This is a partial explanation of why gas giant planets in our solar system orbit beyond the snow line and terrestrial planets orbit inside it. The equilibrium temperature at the snow line in the Barnard's Star system is exceedingly low by terrestrial standards, at about minus 170 °C. The estimated minimum mass of Columbia, plus its location near the snow line, tells us a little about its hypothesized structure. Radial velocity measurements gave a minimum mass of 3.2 Earth masses, so the planet is either a super-Earth or sub-Neptune. Since it is close to the snow line, there is a reasonable probability that the solid part of the planet contains some ices as well as rocks. As it is distant from Barnard's Star, it would have a good chance of having retained its so-called "primordial" atmosphere, a blanket of hydrogen and helium left over from the formation of the system. Earth once had one but it was driven off by the increasing brightness of the Sun early in Earth's evolution [6].

Smaller planets typically lose their primordial atmospheres, and larger ones tend to retain them. A statistical model of this relationship was constructed from Kepler data by Angie Wolfgang and Eric Lopez of UC Santa Cruz, in a well-cited paper published in 2015 [7]. It is now well understood that the fraction of gas in a sub-Neptune planet is directly related to its radius: more gas means a bigger radius. In contrast, planets with more rock have a smaller radius. The Kepler spacecraft, since it measures transits, can measure the planetary radius. Thus, in principle, Kepler data can be used to measure the gas content of planets. However, the relationships between these parameters are not known precisely. Wolfgang and Lopez used a statistical method that accounts for this uncertainty, while using the Kepler data to constrain these statistical relationships. The method involves starting the analysis with our physical understanding of, for example, the relationship between gas content and radius, and then using the Kepler data to tweak these initial assumptions to obtain a final estimate of the actual mass/radius relationship. The upshot of all of this analysis is that a surprising number of super-Earth and sub-Neptune planets might retain at least a portion of their hydrogen and helium primordial atmospheres, even planets that receive considerably more radiation than Earth.

Both calculations and observations suggest that most planets with a radius less than about 1.4 times that of Earth must have lost their original hydrogen/helium envelope, as otherwise they would be larger [8]. Nevertheless, they could then have developed a secondary atmosphere from outgassing, the release of gases from the interior of a planet into its atmosphere, largely by volcanic eruptions [9]. For planets that are large enough to retain a primordial atmosphere, there can be some curious results. Take for instance the planets of Kepler-79, a system that is a mighty 1000 parsecs away in the Vega sector, within the Kepler field of view (see Chap. 3). Since this planet is very far away, it is nowhere near our patch of the known galaxy, but it is such a good illustration of this point that I'll include it anyway. There are both transit and radial velocity measurements for the four known planets of this system, so we know both their radii and their masses. That also means that we know their densities, and this can tell us a lot about the composition of a planet. Planet b has a density a little less than Uranus and has a slightly smaller mass, so this strongly suggests that like Uranus it is a sub-Jovian or sub-Neptune planet with a massive atmosphere. Planet d is smaller (about five to six times the mass that of Earth) but it has an absurdly low density only one-tenth that of water [10]. Known as a "super-puff" planet, this world must have a hydrogen-helium envelope that is some tens of percent of its total mass, very much more than typical for a planet of this mass. It is not known for sure how these super-puffs ended up with so much gas, but one theory is that they collected it when they were much further away from their parent star than they are now, and then later they migrated inwards [11]. Another possibility is that the radius measured during the transit of the planet is actually instead the radius of a giant rocky ring system circling the planet. This would mean that the real radius of the planet is a lot smaller and that the planet is more dense as a result [12]. Yet another possibility is that the measured radius of super-puffs is actually the extent of a high-altitude haze layer [13].

Using results like these to gain insight into the atmospheric composition of a planet like Columbia is problematic, because the Kepler data are utterly dominated by planets that are close to their stars and receive more radiation than Earth does. This is not surprising, as planets that are close to a star are much more likely to transit and therefore be observed by Kepler. Columbia is a lot further out from its star and receives much less radiation than the planets in the Kepler sample, only about 2% of the amount received by Earth. Some relevant information can be gleaned from the Kepler data, though. For planets like Columbia that are most likely in the range of about 3–5 Earth masses, the Kepler sample shows that planets with incoming radiation values similar to Earth may have radii anywhere from about 1.5 to about 3 times that of Earth. This would imply an atmospheric mass ranging from about 0.1 to 2%

of the total mass of the planet, with an average of perhaps about 0.5%. As Columbia receives considerably less incoming radiation than Earth, its actual atmospheric mass percentage could easily be larger than this. This would imply that Columbia would have a very large atmosphere compared to Earth's, as Earth's atmosphere only comprises about 0.0001% of its mass.

We therefore can speculate that the typical atmosphere of a planet like Columbia is a thick one of hydrogen and helium, with a surface pressure of around 5000 times the atmospheric pressure at the surface of the Earth. The intense greenhouse effect of such an atmosphere should guarantee surface temperatures of several hundred degrees C, with the surface itself being a hot ocean, with water the most likely major constituent [14]. Due to the radial velocity technique only giving a minimum mass, though, there is a slight chance that Columbia's mass would be large enough for it to be an ice giant like Uranus but a bit smaller, in other words a sub-Neptune. If so, and if it actually exists, it could be spotted by the JWST. This direct imaging of extra-solar planets will become more common as new observing systems come online.

Analysis of the Kepler transit data indicates that many M stars have super-Earths orbiting close to them. This does not appear to be the case for the Barnard's Star system, as analysis of the radial velocity data has ruled out any planets larger than a minimum mass of 0.7 Earth masses that orbit in 10 days. This analysis also ruled out planets larger than a minimum mass of 1.2 masses that orbit in 40 days [4]. These orbital periods are not chosen at random, as they represent one estimate of the inner and outer boundaries of the habitable zone. Still, a planet with a minimum mass of 0.7 Earth masses in the middle of the habitable zone, with an actual mass of 1.0 Earth masses, is not excluded by these measurements. Such a hypothetical planet could have suffered significant atmosphere loss from the radiation of Barnard's star, a similar situation to planets in the habitable zones of other M stars. One estimate suggests that this planet could have lost about 90 Earth atmospheres of gas over a period of a billion years [15]. But if, like Columbia, it started out with 5000 Earth atmospheres of gas, there might still be plenty left over for the planet to retain a thick atmosphere even billions of years after it formed.

As mentioned earlier, in 2021, new work suggested that the original detection of Columbia was actually a signal from Barnard's Star itself, with the period of the signal related to the slow rotation rate of the M star [5]. In 2024, a smaller planet was found in this system. Luckily for us, though, there is another M star in this sector, Gliese 625, that actually does have a confirmed super-Earth planet. So many of the arguments above that have been made about the apparently imaginary Columbia could apply to the very real planet

Gliese 625 b, so much so that we will call this planet Columbia Mark 2. A big difference is that Mark 2 is a hot super-Earth instead of a cold one like Columbia, but results from the Kepler data suggests that high temperatures are not a big barrier in themselves to the retention of a thick atmosphere [16].

This begs the question: how come that didn't happen in our solar system? How is it that Earth did not retain its hydrogen-dominated atmosphere when plenty of worlds orbiting M stars apparently did? The mass of the Earth is towards the low end of the mass range of the Kepler sample, so that would certainly assist the loss of Earth's primordial atmosphere. But the answer appears to be that the terrestrial planets in the solar system only became fully formed after the hydrogen gas was dispersed from the inner solar system by the growing radiation of the young Sun [17]. It is still not fully understood why some Earth-sized planets managed to retain a primordial atmosphere while planets like Earth did not [16].

Columbia is not the first planet of Barnard's Star to have disappeared after further analysis of the data. In the 1960s, the respected American-based astronomer Peter van der Kamp used the telescope at the Sproul Observatory at Swarthmore College to make precision measurements of the position of Barnard's Star. These apparently showed a side-to-side wobble of the kind that might be caused by the tug of an orbiting planet. Later work showed that the wobble was due to the periodic maintenance of the telescope, and as a result, lots of other stars showed the same wobble. The difficulty was that once van der Kamp was confronted with this evidence by other astronomers who used the Sproul telescope, he refused to believe it. He stuck to his guns despite being proved wrong. Scientists really are human after all.

Barnard's Star b (Unconfirmed) *How similar to Earth: Not very. Super-Earth that likely has a thick atmosphere. But first it has to exist.*
Plausible planet: Thick hydrogen/helium atmosphere with multiple cloud layers (with the lowest one being water clouds). Hot ocean surface with atmospheric pressure several thousand times that of Earth.
Best case scenario: Thinner atmosphere but with sufficient greenhouse effect for more temperate surface conditions. Surface pressure still high.
Worst case scenario: It may not exist!
Cultural connection: Hyperion and Rocheworld.

* * *

A recent development in the science of extrasolar planets is the ability to take observations of extrasolar planetary atmospheres. A case in point is Gliese 486

scenario. At present, though, we have almost no relevant observations to validate the differing predictions of theory and calculations.

There is no guarantee that even if a large moon were to form in orbit around Gliese 687 b that it would necessarily stay there. It might eventually escape or it might end up colliding with the planet it orbits. Recent calculations suggest that these scenarios are particularly prevalent for moons that are satellites of planets orbiting M stars [24]. The ultimate fate of these moons relies on tidal forces. As the tides soak up energy, they change the orbit of the satellite. Due to these forces, for moons that orbit outside the "synchronous radius", the distance from the planet where the orbital period is the same as the planet's day length, the tidal bulges in the planet move faster than the moons. The bulges then push on these moons and would cause their orbits to gradually drift outwards. Eventually, this could lead to the gravitational influence of the M star becoming so significant that it would cause these moons to escape from their orbits. In contrast, moons that orbit inside the synchronous radius are moving faster than the tidal bulges on the planet, which then pull back on the moon, slowing it down and causing it to move closer to the planet. Such moons could eventually spiral inwards, ultimately getting so close to the planet that they risk being torn apart by extreme tidal forces. All of these changes in orbits would take time, and the question is whether that time is large or small compared to the age of the stellar system, and so how long the moon might remain in a stable orbit.

The relevant physical equations to perform these calculations are at present not well constrained. In particular, the rate of dissipation of tidal energy inside planets is poorly known, and this is a crucial factor. Nevertheless, calculations have been performed, indicating that the chance of a moon surviving in orbit around a planet for billions of years is poor for planets that orbit their stars in less than 10 days [30]. Gliese 687 b takes about 38 days to orbit its star, and at this orbital period, survival rates appear to be about 20%, with a large spread of possibilities. For planets with long orbits, more than 100 days, the survival rates are considerably better. All in all, when combined with the low probability of a hypothetical moon of Gliese 687 b being large enough to be habitable, this evidence strongly suggests that even though Gliese 687 b is likely in the habitable zone, it is very unlikely to have a habitable moon.

One slight problem with this scenario is that this system is supposed to host the habitable planet Haven, as described in the *Revelation Space* (2000) series of books by Alastair Reynolds [31]. This highly entertaining space opera series spans multiple millennia and stellar systems, so it is really unfortunate that Haven likely cannot exist, just because Gliese 687 b is in the way. There is also another world in the Gliese 687 system, a cold sub-Jovian that receives about as much radiation as Jupiter does in our solar system. This planet also has a high eccentricity orbit, ranging from about 0.7 to about 1.6 a.u. from its star, meaning that it would prevent other planets forming in its vicinity.

Gliese 687 is in the constellation Draco, the dragon. In one version of the mythology surrounding this constellation, Draco guarded the golden apples of the Hesperides, nymphs who tended a garden traditionally located in the far west of Europe [32]. Hercules was assigned the task of stealing these apples, as one of his famous twelve labors. Now leaving aside the perfectly reasonable legal argument that the golden apples were clearly the property of the Hesperides and that Hercules had no right to steal them, here we are talking about ancient times and different attitudes—the kind of mindset where Homer could casually mention that one of the first acts of Odysseus's long journey back home from the Trojan War was to put the city of Ismarus to the sword, just because he could. After Odysseus and his crew killed all the men of Ismarus, the women of the city were divided amongst Odysseus's men, equally, "so that none might have reason to complain" [33]—apart from the complaints of the women, of course. The term "hero" seems to have been used differently by the ancient Greeks, as it did not necessarily mean someone who was both heroic in deed and who also had good motivations. Hercules himself was certainly heroic but not always well motivated. In fact, he was assigned his twelve labors as punishment for killing his own children.

Whether Hercules really existed as an individual or was an exaggerated combination of many such heroes is unknown. Some legends put his death at about 1226 B.C.E., or a couple of generations before the Trojan War [34]. After his death, Hercules was eventually worshipped as a god, at least in some locations in Greece. This transition from earthly hero to supernatural being can be seen even in our times. For instance, there is a cult that worships Elvis Presley as a god [35]. Add that to his profound cultural impact and the numerous "sightings" of Elvis and he is well on his way to Olympus.

Gliese 687 b *How similar to Earth: A sub-Jovian but in the habitable zone. Only its moon could be habitable, if it exists.*
Plausible planet: Largish exomoon but may not be large enough to be habitable.
Best case scenario: As for "plausible planet".
Worst case scenario: No exomoon, as tidal forces have removed it.
Cultural connection: Hercules and Haven.

*   *   *

We now move on to Hercules's own constellation, where separated by about 1.4 parsecs are the stars GJ 686 and Mu Herculis. The M star GJ 686 hosts a hot super-Earth, but little is known about it. About 8 parsecs away from Earth, Mu Herculis is a quadruple star system, including a G-type sub-giant (Mu Herculis A) and three M stars. One of the M stars orbits Mu Herculis A at an average distance of about 20 a.u., while the other two M stars are about 280 a.u. away and orbit each other with a mean distance between them of about 10 a.u. No planets have yet been discovered orbiting anywhere in this menagerie. The main claim to fame of Mu Herculis A is that due to its brightness, it is one of the best-observed G-type stars. As an evolving sub-giant and former main sequence star, it is old, about 7.8 billion years. A lot of information about this star can be obtained from a technique known as asteroseismology. Like seismic waves in Earth's interior caused by earthquakes, stars have oscillations whose typical frequencies are determined by their internal structure. These frequencies can be deduced from subtle variations in the light output of the star. When combined with numerical modeling, these data can provide an amazing amount of detail. In the case of Mu Herculis A, its mass, gravity, radius, age, and even aspects of its composition can be determined [36]. Similar observations of other nearby Sun-like stars will give much greater precision in estimates of these parameters, important data for our understanding of the evolution of stellar systems and ultimately the habitability of their planets.

Also worth a very quick visit is the Sun-like star Beta Canum Venaticorum, about 8.5 parsecs away, also known as Chara. While no planets have yet been discovered circling it, its system has long been identified as a priority target for exploration, due to Chara's proximity and similarity to our Sun. Radial velocity measurements show that there are likely no giant planets in the system closer than 10 a.u. to the star [37]. Therefore there is room in the system for terrestrial planets in the habitable zone, if they exist, as simulations have shown that they are more likely to form in systems like this one that do not have such giant planets [38].

Noteworthy is another nearby Sun-like star in this sector, Beta Comae Berenices. With no known planets and slightly younger than the Sun, it is still a top candidate system to host a habitable world. It is noteworthy that the constellation Coma Berenices is the only one named after an historical person. The name means "Berenice's hair" and honors Queen Berenice II of Egypt, a Hellenistic ruler of the third century B.C.E. who sacrificed her hair to the gods to ensure that her husband Ptolemy III returned safely from war. Unfortunately, this votive offering then went missing. The court astronomer Conon of Samos diplomatically claimed that the goddess Aphrodite stole it and put it into the sky as a constellation. For some reason, this tall tale was widely accepted at court and elsewhere in Ptolemaic Egypt [39].

Further out, we arrive at Arcturus, both the brightest star in this sector and one of the nearest giant stars. At only 11 parsecs away, it has about the same mass as our Sun but is about 170 times as luminous [40]. Its name means "guardian of the bear" and refers to its position in the sky close to the Big Dipper of Ursa Major, the greater bear. Arcturus itself is in the constellation Boötes, the "herdsman" or "plowman", also referring to its position close to Ursa Major, which the ancient Greeks saw as a cart with oxen [41]. In traditional Chinese astronomy, Arcturus was of considerable importance, as it is the Horn of the Azure Dragon, one of the main Chinese constellations. This dragon is an important symbol throughout east Asia. He shows up as a statue in the magnificent Kiyomizu ("pure water") Temple in Kyoto, set on a giant hillside terrace overlooking lush woods. The dragon is one of the four guardians of Kyoto and protects the city on its eastern side. At night, he is reputed to spring to life and drink from a nearby waterfall. The shrine also has a reputation for its matchmaking and wish-granting powers, hence the large numbers of young people who flock there every year.

Arcturus also played a part in the daring voyages that the Polynesians made across the Pacific hundreds of years ago. During the Northern Hemisphere summer, because the Earth's axis is tilted about 23° to the plane of its orbit around the Sun, stars at declinations of 23° are directly overhead at latitude 23° at midnight during mid-summer. Figure 4.1 shows that the declination of Arcturus is about 19°. The latitude of the Big Island of Hawaii is about 19°, and during mid-summer at this latitude, stars at declination 19° are directly overhead at midnight—like Arcturus. Thus sailors travelling due north from Tahiti during the Northern Hemisphere summer just kept going until Arcturus was directly overhead at midnight, and then they turned west and sailed with the prevailing trade winds until they hit the Big Island [42]. Even so, it is still not completely clear how Polynesian sailors ever found the Hawaiian Islands in the first place, since Hawaii is in the middle of the North Pacific surrounded by hundreds of kilometers of open ocean in all directions. What is clear is that a very high level of sea-faring expertise was needed to cross these empty expanses of ocean, and that celestial navigation played a key role.

Arcturus is not known to have any real planets, but the fictional planet Tormance circling Arcturus is the setting of David Lindsay's stunningly original novel *A Voyage to Arcturus* (1920) [43]. This tale combines elements of science fiction, fantasy and horror with vivid descriptions of colorful alien lifeforms and landscapes. The planet hosts crystal trees, green snow, airborne edible jellyfish, and oceans that are dense enough to walk on. The main character in the book undergoes a number of bizarre, rather depressing adventures during a quest of uncertain purpose that is combined with much philosophical speculation, a kind of *Pilgrim's Progress* without the progress [44]. During the middle of the day on the planet Tormance, the heat of Arcturus is said to be so powerful that it becomes unbearable, which suggests a planet considerably warmer than Earth.

Arcturus is about 7 billion years old and gives us an idea of what our solar system might be like when our Sun becomes a red giant. Currently, for a planet of Arcturus to receive the same amount of stellar radiation that the Earth receives from the Sun, it would have to be about 13 a.u. from Arcturus. This is a bit more than the distance of Saturn from the Sun and begs the question of how the moons of Saturn might adapt to this kind of level of incoming sunlight when the Sun becomes a red giant as luminous as Arcturus, about 5 billion years from now [45]. Titan is the largest of Saturn's moons and is the only solar system moon to have an atmosphere thicker than Earth's. Strangely enough, this combination of thick atmosphere and weak gravity means that on Titan, it would be possible for a fit person to strap on a pair of wings and fly [46]. Simulations indicate that during the Sun's red giant phase, there may well be a period of time lasting as much as a few hundred million years when liquid water oceans would be present on Titan [47]. Detailed climate model simulations of surface conditions on Titan during the Sun's red giant phase have not been performed, however. Thus it is difficult to say what the current climate conditions might be on any hypothetical moons of the equally hypothetical planets currently orbiting Arcturus at about the distance of Saturn from the Sun. Even if the climates on such moons were mild, such conditions would be only temporary, as during the first part of the red giant phase stellar luminosity increases relatively rapidly. Still, such luminosity increases would only be roughly 1% per million years [48]. For human beings, a few million years of mild temperatures on one of these worlds sounds plenty long enough.

Further Afield

Just outside the known galaxy in this sector is the well-studied planet Gliese 1214 b, or Enaiposha as it is now officially called. A number of stars and planets are now being named through suggestions provided by the public [49]. According to the International Astronomical Union (IAU), the official body in charge of star names, "Enaiposha" is a word in the Maa (Masai) language of Kenya, meaning rough waters.

This is perfectly appropriate for this planet. The hot sub-Neptune Enaiposha is a prime candidate to be a waterworld. Here we do not just mean a planet that has lots of water on its surface, but one where water is a significant constituent of its total mass. A special name is probably needed for such worlds, to distinguish them from rocky terrestrial planets that host world-covering oceans of various depths, but whose interiors are still overwhelmingly made of rock. Donald Glaser of the University of Arizona suggests using "pelagic planet" for planets with world oceans, and "water world" for planets more than 1% water by weight, with oceans deep enough to form an ice layer between the water and the crust [50]. Geologically, the two types of worlds would likely have considerably different characteristics and evolution [51]. Estimates of Earth's water mass content vary, due to the uncertainty about how much is inside the Earth in addition to the oceans, but typical values are in the range of 0.03% to about 0.1% of its total mass [50, 52]. But simulations of planetary formation have routinely generated planets with hundreds of times Earth's water content, giving a water content of more than 10%. The dividing line between the two types of watery worlds might be somewhere in the region of 10–100 times Earth's water content, but this is quite speculative, and there are different opinions on how that might affect a planet's climate [51, 53].

Recent results from the JWST have thrown some light on the possible atmospheric composition of Enaiposha [54]. Its atmosphere is likely hazy and comprised mostly of molecules heavier than hydrogen and helium, like carbon monoxide, carbon dioxide, sulfur compounds, and maybe hydrocarbons like methane and ethane as well. Water vapor has also likely been detected. The planet's temperature has also been measured by the JWST, giving a day side temperature of 553 K and a night side temperature of 437 K [55]. The warm night side temperature strongly implies a thick atmosphere, one that is able to transport a substantial amount of heat to the night side. In addition, these measured temperatures are less than the planet's equilibrium temperature if zero albedo is assumed. The planet therefore likely has a reasonably high albedo, most likely around 0.5, or slightly less than that of Venus and about the same as Jupiter. An albedo this high implies the presence of either cloud or haze. Jupiter's moon Titan also has a hazy atmosphere that contains a substantial amount of methane. Unfortunately, the type of haze that forms in Titan's atmosphere has an albedo that is too low to cause the haze on Enaiposha, as Titan's albedo is only about 0.22. Thus it is still not clear what kind of haze blankets Enaiposha. There are a number of candidates at present, but it is too early to decide amongst them. Also, Enaiposha's high albedo differs from the predictions of Sudarsky's theoretical model of gas giant albedos, which suggests a low albedo in this temperature range [56]. The authors of that study do note, however, that the albedo of this type of gas giant could be changed substantially by differences in atmospheric composition or even the presence of thin, high clouds.

While it is clear that Enaiposha has a substantial atmosphere and its possible constituents have been identified, it is not precisely clear yet what the planet itself consists of. As of 2024, there are two main possibilities for the composition of Enaiposha, given its mass and measured radius [55]. It could be a planet with a substantial hydrogen-dominated atmosphere but with considerable heavier molecules as well. A sub-Neptune like Enaiposha that is relatively close to its star might still have retained some of its primordial hydrogen and helium atmosphere, but also might have lost some of it. Since heavier molecules would be lost less easily, the fraction of heavier molecules in its atmosphere would therefore increase over time, thereby explaining their current presence in the planet's atmosphere. Alternatively, the planet might have an atmosphere dominated by heavier molecules but with an interior mostly composed of water. More observations are required before a definitive conclusion can be reached.

Further out in this sector at 31 parsecs is the K-type giant Iota Draconis. Orbiting it is the hot super-Jovian world Hypatia, although at 16 Jupiter masses Hypatia is well into brown dwarf territory and could be one. In 2002, this was the first planet found to orbit a giant star [57]. Its orbit is highly eccentric and it is not in the habitable zone, as it has an equilibrium temperature of about 600 K. There is another cold Jovian or brown dwarf considerably further out in this system, but neither planet is likely to host a habitable world, and no other planets are known at the time of writing.

Hypatia is named after one of the most prominent philosophers of Plato's school in Alexandria (Fig. 4.2). She lived during late Roman times in the first half of the fifth century, when the Egyptian coastal city was still a renowned center of learning. Hypatia was a mathematician and is credited with revising and improving Ptolemy's *Almagest*, among other contributions [58]. As well as being clever, she was reputedly exceedingly beautiful [59]. The period when she lived was one of great turmoil. Only a few years earlier, in 410, the mostly Christian Visigoths had sacked mostly Christian Rome, partly as a result of a long and bitter conflict over many decades, partly because many of the Visigoth soldiers in the Roman army and their families had recently been massacred by the Romans [60]. Under the circumstances, the sack was remarkably restrained, as there was no wholesale massacre of the population, and important monuments like St. Peter's Basilica were left intact.

Fig. 4.2 Fictional portrait of Hypatia (1908) by Jules Maurice Gaspard. Reprinted from Wikipedia (https://commons.wikimedia.org/wiki/File:Hypatia_portrait.png, accessed on the 8th of July 2024). © 1908, Jules Maurice Gaspard. Public domain

While much of the population of the remaining portions of the Roman Empire had converted to Christianity (like the Visigoths), the conflict between non-Christians and Christians was still ongoing, and Alexandria was no exception [61]. There was an open power struggle between Orestes, the secular Roman leader, and Cyril, the Christian patriarch of Alexandria. Riots broke out and both sides committed atrocities. Hypatia was an advisor to Orestes and widely respected in the city, putting her in opposition to Cyril. In 415, after one of Cyril's priests was tortured to death by Orestes in retaliation for a riot that almost cost Orestes his life, a Christian mob dragged Hypatia from her carriage and flayed her alive using shards of pottery or oyster shells. It was widely believed in Alexandria that Cyril had ordered her murder. Moreover, the assassins escaped punishment, and with the death of Hypatia, Orestes lost a politically powerful supporter. He left Alexandria, leaving Cyril as the city's dominant political figure. Cyril lived to a fine old age and later was made a saint [62].

At 32 parsecs away in this sector, we encounter the K-star HD 110067, whose system hosts six close orbiting, hot sub-Neptune worlds whose orbital periods are multiples of each other. This resonant orbital arrangement is not unknown but one with six planets is rare [63]. The TRAPPIST-1 system (Chap. 9) is an example of a system that is almost in resonance, but those planets circle a very faint star. In contrast, HD 110067 is a robust early K star with a luminosity of at least 40% of our Sun, making it a lot easier to extract atmospheric spectra from its planets. The potential for useful observations with the JWST is high [64].

Finally, at 47 parsecs out, we reach Eltanin, or Gamma Draconis. It actually appears brighter in our sky than both Alpha or Beta Draconis, and is a K-type giant about 590 times as luminous as the Sun. For a planet circling it to receive the same amount of stellar radiation as Earth does, it would have to be about 24 a.u. away and would take about 80 years to orbit its star. If a hypothetical Earth-like planet orbited there, it would have seasons lasting many years, and naturally this was part of the plot of Ursula Le Guin's *Planet of Exile* (1966), set in this system [65]. In this novel, the Earth-like planet Werel orbits a large "moon" of unspecified size, probably a giant planet, taking about 400 days to circle it. In turn, the "moon" orbits Eltanin in sixty of these 400-day "moonphases", in total about 66 years. The novel is about a small settlement of Earthlings who have been stranded on Werel for several hundred years, cut off from contact with the rest of humanity. Their interactions with the local indigenous people are the main conflict in the story. Like all of Le Guin's work, the novel is full of highly imaginative detail, and the author skilfully creates an atmosphere of mystery, tragedy and awe. The book is set in the same universe as the author's acclaimed, prize-winning novel *The Left Hand of Darkness* (1969) [66]. Both tales are part of the Hainish cycle, named after the race of humanoid extraterrestrials who were said to have colonized Earth and other nearby worlds several hundred thousand years ago.

Could a world like Werel exist? There are a number of issues. The first, naturally, is Eltanin itself. As an evolved K-giant of about 2 solar masses, Eltanin would originally have been a main sequence star of class A, and would have had a main sequence lifetime of about 2 billion years [67]. During that time, its habitable zone would have been centered at roughly 4 a.u. or so from the star, although like the habitable zone of the Sun, it would have gradually expanded outwards as the star evolved along the main sequence. But its K-giant phase will last considerably less than a billion years [48]. During this period, any planets circling at the distance of the K-giant phase habitable zone would have had little time to develop sophisticated life, and during the star's main sequence phase, they would have been outside the zone of high stellar radiation where the creation of an abiotic (i.e. non-biological in origin) oxygen atmosphere by photodissociation would have been possible (see Chap. 3). The other issue is the mutual orbit of Werel and its moon. A period of 400 days may not be impossible but there are no large moons in the outer part of our own solar system with this type of orbital configuration (there are moons of Jupiter and Saturn with similar orbital periods but they are very small) [68, 69]. Formation of such a distant moon in situ, due to gravitational collapse of a dust cloud like the one that formed the moons of Jupiter, would be very unlikely [70]. So capture is the most likely alternative, with the gas-giant moon of Werel having grabbed the smaller planet into a distant orbit sometime in the past. The long-term stability of such a large orbit in the outer stellar system of a massive star could also be in doubt.

Overall, it seems unlikely that Werel could actually exist as described in *Planet of Exile*. But does it really matter? Only a little bit. The Hainish cycle is full of scientific improbabilities, but criticizing it too much on that basis is like carping at Jules Verne because he suggested using a cannon to fire astronauts to the Moon instead of recommending a Saturn V rocket. Read and enjoy.

Arcturus Sector Summary

In the near future, we will be able to reliably detect Earth-size planets in the habitable zones of stars (see Chap. 10). When we can do this, there are a number of candidate systems in this sector to explore, from Barnard's Star to Chara to Beta Comae Berenices. The atmosphere of Enaiposha still needs to be better characterized, along with the environments of the other known exoplanets, such as the six-planet resonant system of HD 110076.

The sky in this sector is dominated by the two large masculine figures of the huntsman Boötes and the folk hero Hercules. The feats of Hercules are renowned, but deeds that might have been praised in the anarchic, sparsely populated world of ancient Europe would lead straight to prison in our present-day, buttoned-down, crowded society. Still, there stands Hercules, prominent in the summer Northern Hemisphere sky, where he will stay for thousands of years. Sometimes might does make right.

References

1. Brittanica (2023) Barnard's Star. https://www.britannica.com/place/Barnards-star. Accessed 12 Sept 2023
2. Simmons D (1989) Hyperion. Doubleday, New York
3. Forward RL (1990) Rocheworld. Simon and Schuster, New York
4. Ribas I, Tuomi M, Reiners A (2018) A candidate super-Earth planet orbiting near the snow line of Barnard's Star. Nature 563(7731):365–368
5. Lubin J, Robertson P, Stefansson G et al (2021) Stellar activity manifesting at a one year alias explains Barnard b as a false positive. Astronom J 162(2):61
6. Lopez ED, Fortney JJ (2013) The role of core mass in controlling evaporation: the Kepler radius distribution and the Kepler-36 density dichotomy. Astrophys J 776(1):2
7. Wolfgang A, Lopez E (2015) How rocky are they? The composition distribution of Kepler's sub-Neptune planet candidates within 0.15 AU. Astrophys J 806(2):183
8. Rogers JG, Gupta A, Owen JE, Schlichting HE (2021) Photoevaporation versus core-powered mass-loss: model comparison with the 3D radius gap. Mon Not Roy Astronom Soc 508(4):5886–5902
9. Moran SE, Stevenson KB, Sing DK (2023) High tide or riptide on the cosmic shoreline? A water-rich atmosphere or stellar contamination for the warm super-Earth GJ 486 b from JWST observations. Astrophys J Lett 948(1):L11
10. Jontof-Hutter D, Lissauer JJ, Rowe JF, Fabrycky DC (2014) Kepler-79's low density planets. Astrophys J 785(1):15
11. Chachan Y, Jontof-Hutter D, Knutson HA et al (2020) A featureless infrared transmission spectrum for the super-puff planet Kepler-79 d. Astronom J 160(5):201
12. Piro AL, Vissapragada S (2020) Exploring whether super-puffs can be explained as ringed exoplanets. Astronom J 159(4):131
13. Gao P, Zhang X (2020) Deflating super-puffs: impact of photochemical hazes on the observed mass–radius relationship of low-mass planets. Astrophys J 890(2):93
14. Koll DD, Cronin TW (2019) Hot hydrogen climates near the inner edge of the habitable zone. Astrophys J 881(2):120
15. France K, Duvvuri G, Egan H et al (2020) The high-energy radiation environment around a 10 Gyr M dwarf: habitable at last? Astronom J 160(5):237
16. Owen JE, Shaikhislamov IF, Lammer H et al (2020) Hydrogen dominated atmospheres on terrestrial mass planets: evidence, origin and evolution. Space Sci Rev 216:1–24
17. Owen JE, Mohanty S (2016) Habitability of terrestrial-mass planets in the HZ of M Dwarfs—I. H/He-dominated atmospheres. Mon Not Roy Astronom Soc 459(4):4088–4108

18. Trifonov T et al (2021) A nearby transiting rocky exoplanet that is suitable for atmospheric investigation. Science 371(6533):1038–1041. https://doi.org/10.1126/science.abd7645
19. Krissansen-Totton J, Thompson M, Galloway ML, Fortney JJ (2022) Understanding planetary context to enable life detection on exoplanets and test the Copernican principle. Nat Astron 6(2):189–198
20. Moran SE, Stevenson KB, Sing DK et al (2023) High tide or riptide on the cosmic shoreline? A water-rich atmosphere or stellar contamination for the warm super-Earth GJ 486b from JWST observations. Astrophys J Lett 948(1):L11
21. Burt J, Vogt SS, Butler RP et al (2014) The Lick–Carnegie exoplanet survey: Gliese 687 b: a Neptune-mass planet orbiting a nearby red dwarf. Astrophys J 789(2):114. https://doi.org/10.1088/0004-637X/789/2/114
22. Hill ML, Kane SR, Duarte ES et al (2018) Exploring Kepler giant planets in the habitable zone. Astrophys J 860(1):67
23. Heller R (2012) Exomoon habitability constrained by energy flux and orbital stability. Astronom Astrophys 545:L8
24. Dobos V, Charnoz S, Pál A et al (2021) Survival of exomoons around exoplanets. Publ Astronom Soc Pac 133(1027):094401
25. Zollinger RR, Armstrong JC, Heller R (2017) Exomoon habitability and tidal evolution in low-mass star systems. Mon Not Roy Astronom Soc 472(1):8–25
26. Heller R, Pudritz R (2015) Conditions for water ice lines and Mars-mass exomoons around accreting super-Jovian planets at 1-20 AU from Sun-like stars. Astronom Astrophys 578:A19
27. Moraes RA, Vieira Neto E (2020) Exploring formation scenarios for the exomoon candidate Kepler 1625 b I. Mon Not Roy Astronom Soc 495:3763–3776. https://doi.org/10.1093/mnras/staa1441
28. Heller R, Hippke M (2023) Large exomoons unlikely around Kepler-1625 b and Kepler-1708 b. Nat Astron 8:193–206. https://doi.org/10.1038/s41550-023-02148-w
29. Rufu R, Canup RM (2017) Triton's evolution with a primordial Neptunian satellite system. Astronom J 154(5):208
30. Ibid. (ref. [24])
31. Reynolds A (2000) Revelation space. Gollancz, London
32. Ridpath I (2018) Star tales. Lutterworth, Cambridge, p 91
33. Hutchins R, Adler M (1952) Homer, the Odyssey. In: Great books of the Western world. Encyclopedia Brittanica, Chicago, IL
34. Wikipedia (2023) Heracles. Accessed 21 Sept 2023
35. Harrison T (2016) The death and resurrection of Elvis Presley. University of Chicago Press, Chicago, IL
36. Li T, Bedding TR, Kjeldsen H et al (2019) Asteroseismic modelling of the subgiant μ Herculis using SONG data: lifting the degeneracy between age and model input parameters. Mon Not Roy Astronom Soc 483(1):780–789

37. Kane SR, Turnbull MC, Fulton BJ et al (2020) Dynamical packing in the habitable zone: the case of Beta CVn. Astronom J 160(2):81
38. Raymond SN, Armitage PJ, Moro-Martin A et al (2012) Debris disks as signposts of terrestrial planet formation—II. Dependence of exoplanet architectures on giant planet and disk properties. Astronom Astrophys 541:A11
39. Ridpath I (2018) Star tales. Lutterworth, Cambridge, p 79
40. Ramírez I, Allende Prieto C (2011) Fundamental parameters and chemical composition of Arcturus. Astrophys J 743(2):135. https://doi.org/10.1088/0004-637X/743/2/135
41. Ridpath I (2018) Star tales. Lutterworth, Cambridge, p 53
42. Wikipedia (2024) Hokule'a. Accessed 25 Jan 2024
43. Lindsay D (1920) A voyage to Arcturus. Penguin, London
44. Hume K (1978) Visionary allegory in David Lindsay's "A voyage to Arcturus". J Engl German Philol 77(1):72–91
45. Schröder K-P, Connon Smith R (2008) Distant future of the Sun and Earth revisited. Mon Not Roy Astronom Soc 386(1):155–163. https://doi.org/10.1111/j.1365-2966.2008.13022.x
46. Zubrin R (2000) Titan. In: Entering space: creating a spacefaring civilization. TarcherPerigee, Los Angeles, pp 163–166
47. Lorenz RD, Lunine JI, McKay CP (1997) Titan under a red giant sun: A new kind of "habitable" moon. Geophys Res Lett 24(22):2905–2908
48. Pols OR, Schröder K-P, Hurley JR et al (1998) Stellar evolution models for Z = 0.0001 to 0.03. Mon Not Roy Astronom Soc 298(2):525. https://doi.org/10.1046/j.1365-8711.1998.01658.x
49. IAU (2022) 2022 approved names. https://www.nameexoworlds.iau.org/2022approved-names. Accessed 10 Feb 2024
50. Glaser DM, Hartnett HE, Desch SJ et al (2020) Detectability of life using oxygen on pelagic planets and water worlds. Astrophys J 893(2):163
51. Kite ES, Ford EB (2018) Habitability of exoplanet waterworlds. Astrophys J 864(1):75
52. Jacobson SA, Walsh KJ (2015) Earth and terrestrial planet formation. In: Badro J, Walter M (eds) The early Earth: Accretion and differentiation. Wiley, Hoboken, NJ, pp 49–70
53. Foley BJ (2015) The role of plate tectonic–climate coupling and exposed land area in the development of habitable climates on rocky planets. Astrophys J 812(1):36
54. Gao P, Piette AA, Steinrueck ME et al (2023) The hazy and metal-rich atmosphere of GJ 1214 b constrained by near and mid-infrared transmission spectroscopy. Astrophys J 951(2):96
55. Kempton EMR, Zhang M, Bean JL et al (2023) A reflective, metal-rich atmosphere for GJ 1214 b from its JWST phase curve. Nature 620:67–71. https://doi.org/10.1038/s41586-023-06159-5

56. Sudarsky D, Burrows A, Pinto P (2000) Albedo and reflection spectra of extrasolar giant planets. Astrophys J 538(2):885

57. Frink S, Mitchell DS, Quirrenbach A et al (2002) Discovery of a substellar companion to the K2 III giant ι Draconis. Astrophys J 576(1):478

58. Ptolemy, Toomer GJ (trans) (1998) Ptolemy's Almagest. Princeton University Press, Princeton, NJ

59. Booth C (2017) Hypatia: mathematician, philosopher, myth. Fonthill Media, London

60. Wikipedia (2024) The sack of Rome. Accessed 20 Jan 2024

61. Wikipedia (2024) Hypatia. Accessed 20 Jan 2024

62. Ibid.

63. Cowing K (2023) Six planets orbit HD110067 system in a harmonic rhythm. https://astrobiology.com/2023/11/six-planets-orbit-hd110067-system-in-a-harmonic-rhythm.html. Accessed 21 Jan 2024

64. Luque R, Osborn HP, Leleu A et al (2023) A resonant sextuplet of sub-Neptunes transiting the bright star HD 110067. Nature 623:932–937. https://doi.org/10.1038/s41586-023-06692-3

65. Le Guin U (1966) Planet of exile. Ace, New York

66. Le Guin U (1969) The left hand of darkness. Ace, New York

67. Hansen CJ, Kawaler SD (1994) Stellar interiors: physical principles, structure, and evolution. Birkhäuser, Basel

68. Wikipedia (2024) Moons of Jupiter. Accessed 23 Jan 2024

69. Wikipedia (2024) Moons of Saturn. Accessed 23 Jan 2024

70. Ronnet T, Johansen A (2020) Formation of moon systems around giant planets—Capture and ablation of planetesimals as foundation for a pebble accretion scenario. Astronom Astrophys 633:A93

5

Rogue Stars and Lava Worlds

The Sirius Sector: 6–12 h, 0 to –90°

Meaning "scorching" in ancient Greek [1], Sirius is the brightest star in the night sky because it is both more luminous than the Sun and also nearby, only 2.6 parsecs away. Due to its brightness, it has been used for millennia in navigation and to mark the passage of the seasons. The Sirius system consists of a main sequence A star, the nearest to Earth of its type, and a white dwarf, also the nearest of its kind. There are many myths, legends and stories about Sirius, but worth mentioning first is the strange association that so many different cultures have made between Sirius and dogs [2]. For the Greeks, it is the brightest star in the constellation Canis Major, the great dog. In China, it is known as the "celestial wolf". In North America, the Cherokee, the Blackfoot and other First Nations peoples associate the star with canines of various kinds. The wide geographical dispersion of this association suggests that the original myth was of great antiquity. This myth is the basis of the expression "dog days", meaning the height of summer in the Northern Hemisphere, when Sirius first appears in the early dawn sky after months of absence [3] (Fig. 5.1).

Sirius is bright but 200 million years ago it would have been even brighter. At that time, the white dwarf Sirius B would have been a main sequence B star and would have been both brighter and more massive than Sirius A. Like all B stars, though, its time on the main sequence was short. The less massive Sirius A will spend longer on the main sequence, about a billion years in total, before becoming a red giant and eventually a white dwarf [2].

As Sirius is both bright and nearby, there are some science fiction stories about it, but perhaps fewer than one might expect. The very earliest work of

© The Author(s), under exclusive license to Springer Nature Switzerland AG 2024
K. J. E. Walsh, *Planets of the Known Galaxy*, Science and Fiction,
https://doi.org/10.1007/978-3-031-68218-6_5

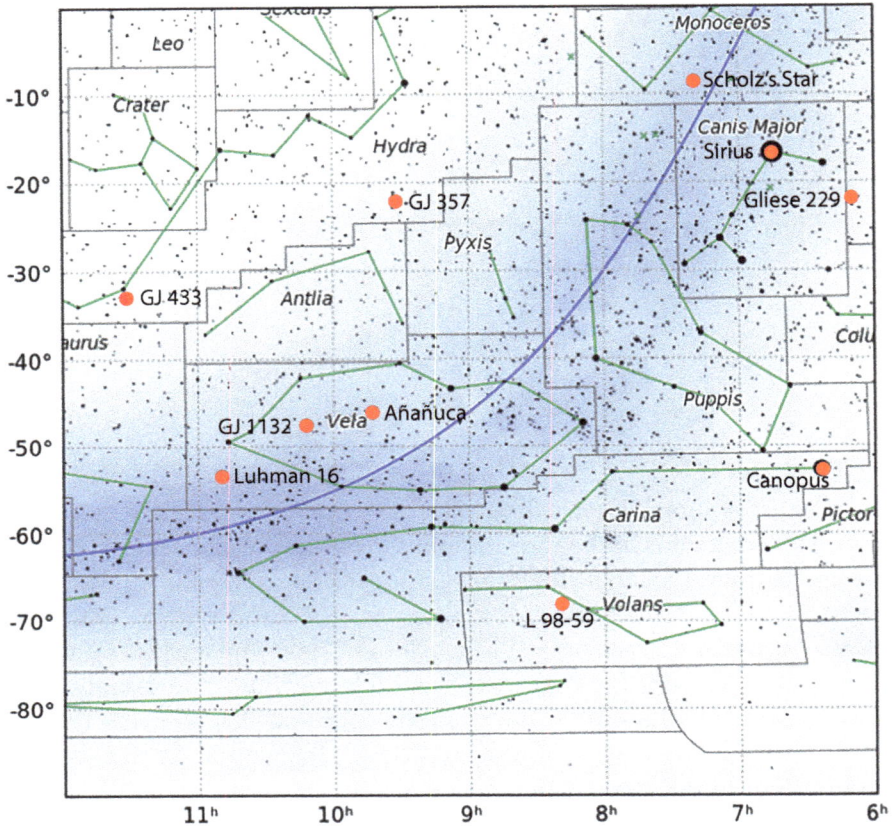

Fig. 5.1 Map of the Sirius sector. Blue shading indicates the "Milky Way", the most star-packed part of the galaxy, while the blue (darker) line is the galactic equator, an imaginary circle running through the Milky Way that slices the galaxy in half. The orange (lighter) line is the path of the Sun's location through the seasons (indicating the "zodiac" constellations). Larger circles indicate brighter stars, while locations mentioned in this chapter are indicated by red circles. Image produced using StarCharter (https://github.com/dcf21/star-charter, accessed on the 2nd of July 2024. © 2020, Dominic Ford) software under the GNU general public license

science fiction is reputed to be the *True History* of Lucian of Samosata, a Syrian author who lived in the second century [4]. In this satire, the author weaves a wild tale of interplanetary travel, full of the most ridiculous events. At one point, some dog-faced men from Sirius make an appearance, riding into battle on "winged acorns". Lucian seems to have been mocking ancient authors who credulously reported the most fantastical tales as truth, perhaps in much the same way that Cervantes wrote *Don Quixote* as a parody of the hackneyed chivalrous romance tales that were popular in his day. Centuries later, Voltaire wrote *Micromegas* (1752) [5], a satire that tells the story of the giant Micromegas of Sirius. He is about 40 km tall and travels to Earth to witness both the cleverness and the stupidity of humanity. The successful TV show *V* (1983–1985) [6] was about aliens from the Sirius system invading Earth, a thinly-veiled allegory of a Nazi takeover.

Another modern cultural link is with the avant-garde composer Karlheinz Stockhausen, who claimed that he came from Sirius [7]. Once you listen to his music, you'll understand why.

Perhaps one of the reasons that there are fewer tales about Sirius than there otherwise might be is because modern writers know that the Sirius system is hostile to human life. Sirius A is hot, young, and has a white dwarf companion an average of only 20 a.u. away. This is not a friendly place for a habitable planet. The evolution of the Sirius system has been full of drama [8]. The whole system is only about 225 million years old, and during that short time, its stars have formed, Sirius B has gone from being a main sequence B star to a giant, has thrown off much of its mass and then has settled down to the white dwarfdom of old age. Any planets that originally formed around Sirius B may not have survived this process. It is not impossible for planets to survive the giant star phase: the giant planet Fortitudo (Chap. 3) comes to mind immediately. Still, giant stars are very bright and emit a strong stellar wind. Not only can this strip the atmosphere of a planet, but if the planet were close enough to its star, it could be entirely engulfed by the star and vaporized. As the star throws off mass, its gravity would become weaker, and this could cause the orbits of its planets to expand, perhaps even making their orbits unstable and ejecting them from the system [9].

Even so, nature has a way of throwing up counter-examples to any theory, and there is at least one giant planet survivor orbiting close to a white dwarf star [10]. The white dwarf WD 1856+534 lives about 25 parsecs away in Draco, but in the Vega sector (Chap. 3; see Fig. 3.1). Its planet is a cold Jovian of several Jupiter masses, although since we only have transit measurements at present but no radial velocity data, this mass is just an estimate. It orbits only 0.02 a.u. from its white dwarf primary, but it cannot have been there during the star's giant phase because it would have been destroyed. Instead, it likely migrated inward since then. How? Usually, when stars lose mass, planets migrate outwards. Here what happened is that as the star lost mass, the planet gravitationally interacted with other bodies in the same system and was thrown into an orbit that closely approached the white dwarf. While such an orbit would most likely have been very eccentric at first, the strong tides raised by the white dwarf on the planet rapidly dissipated its orbital energy. The planet then ended up in the minimum energy state, namely a circular orbit close to the white dwarf.

A prime candidate for the mechanism causing these gravitational interactions is something called the Lidov-Kozai (or LK) effect [11]. This involves a distant body influencing the orbit of two other objects, so that a mutual oscillation is set up between the eccentricity and the inclination of the orbit. A circular orbit can become eccentric, and in response the inclination of a planet's orbit can be modified. In the case of WD 1856+534, the remote perturber of the planet's orbit is a pair of M stars located about 1500 a.u. distant. A plausible scenario here is that the LK effect perturbed the originally distant, circular orbit of the planet into an eccentric orbit with a close approach to the white dwarf, and tidal effects did the rest.

One very curious feature of this system is that the planet is actually considerably larger than its star. The planet has a radius of about ten times that of Earth. In contrast, white dwarf stars are very dense, and WD 1856+534 has about half the mass of our Sun packed into a sphere only slightly larger than Earth. Also, when the planet transits the white dwarf, it only obscures about half of the star. This is very odd considering that the planet is a lot bigger than the star. Much more likely would be the planet completely obscuring the star. While the example of WD 1856+534 shows that it is at least possible for planets to survive until the white dwarf stage, no such planets have yet been discovered circling Sirius B. Given the tumultuous history of the Sirius system, there is no reason for optimism that they do exist there [8]. During the past 200 million years, Sirius B has formed, matured, and grown old, over a period of geologic time that is so short that when our solar system was that age, it was still in stellar kindergarten.

*   *   *

While Sirius has no known planets, in this sector there are many planets elsewhere. The first place we will visit is the Luhman 16 system, the third-closest known stellar system to Earth and the closest known to Alpha Centauri. Strictly speaking, the two components of this system are both brown dwarfs rather than stars. The system is named after Kevin Luhman, an astronomer at Penn State who announced its discovery in 2013 [12]. It has become conventional to name nearby stars after their discovers (e.g. Barnard's Star). An alternative name for the system is WISE J104915.57−531906.1, named after the spacecraft that took the discovery images of the system. Guess which name astronomers prefer.

The two components of the Luhman 16 system are separated by about 3.6 a.u. and are about 600–800 million years old [13]. Age is really important for the discovery and classification of brown dwarfs. Instead of getting brighter with age as main sequence stars do, brown dwarfs become cooler and dimmer, so young ones are brighter and easier to find than old ones.

What would they look like, up close? One useful comparison is to the actual colors of stars as they would appear to a human observer located within their stellar systems. The conventional classification of stars into G-type "yellow", K-type "orange" and M-type "red" stars is very widely used (see Chap. 1) but it is a little misleading. Almost all M-type stars do not actually appear red, when seen close up. This can be deduced by examining the Planckian locus, the dominant color that a hot object emits at a given temperature [14]. An analogy is with heated metal: heat it a little bit and it can glow red, heat it a lot and it can become white-hot. The Sun has a surface temperature of about 5800 K, so is perceived by the human eye as white, as NASA astronauts in orbit have confirmed. When seen from Earth's surface, the Sun looks mostly white but also a little yellow, but that is just because of the reddening effect of the atmosphere. This phenomenon reaches its maximum when the Sun is low in the sky at dawn and dusk, when its light passes through the greatest amount of atmosphere. The largest M stars have temperatures of about 3700 K, and thus will also be perceived as white. The smallest ones have effective temperatures of about 2400 K, and here is where some color might be seen, typically a pale yellowish shade.

To get a better feeling for what these small stars might really look like, take the small M star 2M1540, located in the Alpha Centauri sector (Chap. 2) in the dim constellation Norma about 4.5 parsecs away. Like all of the color-based stellar classifications, the M star category is divided into a number of divisions, ranging from M0, a large M star, to M9, a small star on the boundary of the L-class brown dwarf category (in this classification, our Sun is denoted G2). The star 2M1540 is classified as M7 and is the closest M7 star to Earth [15]. Due to this small star being some 1600 times less luminous than the Sun, a habitable planet circling it would have to huddle only about 3.5 million kilometers from it. At that distance, because most of the radiation given off by 2M1540 is in the infrared, the amount of visible light illumination that it would give out would be about 30 times less than the Sun provides to the surface of the Earth when the sky is clear. This star would thus give about the same level of illumination as in a very well-lit factory, or on a bright overcast day at midday on Earth. Another comparison can be made with our Sun at sunrise, when it can be gazed at directly without discomfort for a short period of time. At that time of day, total levels of illumination are about a quarter of that given by 2M1540. So it is possible that a small M star like 2M1540 would also generally not be painful to look at for a short period of time, and it is also conceivable that it might show its true color, in this case yellowish or perhaps a pale beige [16]. It has to be said, though, that a journalist probably would not make "beige dwarf" their first choice to describe this star.

But back to Luhman-16. Component A is classified L7.5, with a temperature of 1350 K, while component B is T0.5, with a temperature of 1210 K. Both these temperatures are well into the orange-red region of the true-color spectrum. Again, though, there is a wrinkle: the atmospheres of these brown dwarfs would likely interfere with the transmission of visible light in such a way that the dwarfs would appear purple or violet, or possibly magenta. The very coolest Y dwarfs would be black or a dull red [17]. An example of a Y dwarf is WISE 0855-174, in this sector only 2.3 parsecs away. Also discovered by Kevin Luhman, this Y4 dwarf is the coldest object of its type known to be floating around between the stars, with a temperature between 225 and 260 K [12].

We move further out in this sector to another system found by the WISE spacecraft, WISE J0720-0846, otherwise known as Scholz's Star, named after its discoverer Ralf-Dieter Scholz, who works at the Leibniz Institute in Potsdam, Germany. Scholz's Star is a good illustration of the fact that stars move around and do not always stay the same distance from Earth. About 70,000 years ago, it passed within about 50,000 a.u. of the Earth, or through the Oort cloud of comets surrounding our solar system [18]. An even better example of a wandering star is Gliese 710, a K star currently in the constellation Serpens Cauda, at a distance of about 19 parsecs in the Fomalhaut sector (see Chap. 9). It is reasonably confidently predicted that about 1.3 million years from now, it will pass by our solar system at the nail-biting distance of only about 10,000 a.u [19].

This distance is well inside the Oort cloud and has the potential to disrupt the orbits of many of the comets there. How much this would affect the comet infall rate into our solar system is uncertain, but there is the potential for this rate to greatly increase as the star begins to throw its weight around and toss comets in all directions. A large comet infall rate would be damaging to the development of life on habitable worlds. The exact role of comets in mass extinction events on Earth remains controversial [20], but it goes without saying that more comets would mean a higher chance of such extinction events. As Gliese 710 slowly passes by, it will become by far the brightest star in the sky, about equal to Jupiter at its brightest.

While such close passes to our solar system by other stars are not frequent, they are frequent enough to ask the question how often a star might approach our solar system, or other systems, at a distance close enough to be really disruptive to the actual orbits of the planets in that system. Luckily, this is rare in our sparsely starred region of the Milky Way galaxy. For our solar system, about once in a billion years we might expect a star to travel within about 1500 a.u [21]. The chance that during Earth's entire four and a half billion year history a star might come as close to the Sun as Earth, and thereby risk ejecting our planet into interstellar space, is less than one in a billion. There is an old science fiction movie called *When Worlds Collide* (1951), about the destruction of Earth by a passing rogue star, so this is not a new idea [22].

But there are upwards of 100 billion stars in the Milky Way. This greatly raises the odds that somewhere out there, destruction or ejection is exactly what has happened to more than one unlucky terrestrial planet. Let's just hope there was no one living on it at the time.

*   *   *

The first planetary system that we visit in this sector is that of GJ 433, an M1 star about 9 parsecs distant in the constellation Hydra. The largest of the constellations, Hydra sprawls across almost half the sky from east to west but is thin from north to south. In Babylonian and later Greek mythology, its long, thin shape represents a serpent. Hydra was also a monster with many heads. In the second labor of Hercules, he traveled to the monster's lair to slay it [23]. He tried cutting off its heads, but this did not work because more heads grew in their place. With the help of his sidekick, Hercules hit upon the idea of cauterizing the wounds made by cutting off Hydra's heads so that they could not grow back. The myth of Hydra has entered into modern English to describe a problem that firmly resists all attempts to solve it and keeps on cropping up (Fig. 5.2).

GJ 433 is in the far southern part of Hydra. Table 5.1 shows that the most temperate of its three planets is planet d. Only discovered in 2020, its climate has not yet been simulated [24]. Being on the borderline between the super-Earth and sub-Neptune classes, its habitability is probably also borderline, and much more work is needed to characterize its atmosphere and surface conditions. Further out in this system is the very cold sub-Jovian GJ 433 c. Receiving slightly less total stellar radiation than Uranus, GJ 433 c is an excellent candidate for direct imaging due to its size and large separation from its star [25]. A lot of information could be obtained if it were to be imaged directly, and this will no doubt be attempted soon.

Another of the many Gliese stars in this sector is the very inactive M2.5 star GJ 357, also located in Hydra. That the star is inactive is good, as it means that it no longer suffers from flares that can damage the atmospheres of its three known planets (Table 5.2). For the innermost planet, we have both

Fig. 5.2 Ercole e l'Idra (Hercules and the Hydra). Reprinted from Wikipedia (https://commons.wikimedia.org/wiki/File:Antonio_del_Pollaiolo_-_Ercole_e_l%27Idra_e_Ercole_e_Anteo_-_Google_Art_Project.jpg, accessed on the 8th of July 2024). © 1475, Antonio del Pollaiuolo. Public domain

Table 5.1 Planets of Gliese 433

Planet	Mass (Earth = 1)	Orbital period (days)	Distance from star (Earth distance from Sun = 1)	Equilibrium temperature (degrees K) (Earth = 255 K)	Type
GJ 433 b	>5.9	7.4	0.062	439–480	Hot super-Earth or sub-Neptune
GJ 433 d	>4.9	36.1	0.178	259–283	Warm or hot super-Earth or sub-Neptune
GJ 433 c	>28.7	4874	4.69	50–55	Very cold sub-Jovian

Table 5.2 Planets of Gliese 357

Planet	Mass (Earth = 1)	Orbital period (days)	Distance from star (Earth distance from Sun = 1)	Equilibrium temperature (degrees K) (Earth = 255 K)	Type
GJ 357 b	2.08	3.9	0.033	482–527	Hot super-Earth
GJ 357 c	>3.67	9.1	0.061	355–389	Hot super-Earth
GJ 357 d	>6.1 or >7.2	55.7	0.204	194–212	Warm super-Earth or sub-Neptune

radial velocity and transit measurements, so we know its mass and radius. These give a planetary density similar to Earth, so this is a world largely composed of rock. For planets c and d, we only have radial velocity measurements. It seems reasonable, though, that these planets would have approximately the same orbital inclination as the innermost one, given that stellar systems form from a flat pancake of gas and dust, all orbiting in roughly the same plane. If that is the case, since the orbit of the innermost planet is known to be almost edge on, then the actual masses of the outer planets are very close to their minimum masses given in Table 5.2.

Discovered by the Transiting Exoplanet Survey Satellite (TESS) team in 2019, planet d has an equilibrium temperature considerably lower than Earth's [26]. As we have seen, a thick atmosphere could make its surface conditions considerably warmer than its equilibrium temperature. Its potential habitability has been given a preliminary assessment in a study led by Lisa Kaltenegger at Cornell [27]. If planet d is largely rocky, it would have a radius of about 1.75 times that of Earth, and if it is a water world, it would have a radius of about 2.4 times Earth's. The authors use a simple climate model to estimate the surface temperature of the planet for a number of different possible atmospheres. As this planet receives less incoming radiation than Mars, if an Earth-like atmosphere is assumed the mean surface temperature is naturally well below freezing, about 210–215 K, or about −60 °C (−80 °F), or about the same as Mars. Such a large world could easily have an atmosphere with a fair amount of carbon dioxide in it. The authors also simulated the climate if the concentration of carbon dioxide is increased to about one-tenth of the amount of the planet's atmosphere. In that case, mean surface temperatures are slightly above freezing, and if the planet had an atmosphere with double the atmospheric pressure of Earth and if 10% of it were carbon dioxide, mean temperatures would be similar to Earth's. Call it Greenhouse World.

Planets like Greenhouse World require large amounts of carbon dioxide in their atmospheres to maintain habitable temperatures, to compensate for receiving less incoming stellar radiation. They are probably reasonably common, as the simulations of planetary atmospheric evolution by Joshua Krissansen-Totton (see Chap. 4) generate a number of them. One significant issue for the human habitability of such worlds is that the amount of carbon dioxide required to keep their climates warm is also unbreathable and toxic to human beings. A 10% carbon dioxide atmosphere may be deadly for humans, no matter how much oxygen it also contains [28]. Still, a smaller version of Greenhouse World, a planet with a surface gravity similar to Earth's, an atmosphere that is not too thick and a temperate climate, would be very hospitable to life in general and even not particularly inhospitable to human life.

Gliese 357 d *How similar to Earth: Larger. On the borderline between super-Earth and sub-Neptune.*
Plausible planet: Large, cold world with thick atmosphere.
Best case scenario: Due to strong greenhouse effect, Earth-like temperatures. Planet is large and will probably have strong surface gravity.
Worst case scenario: An uninhabitable sub-Neptune.
Cultural connection: Greenhouse World.

*   *   *

Lava worlds are planets where the surface temperature is so high that oceans of molten rock and metal have been created, like the lava flowing out of a volcano on Earth but covering much of the planet [29]. There is at least one lava world in this sector, namely GJ 367 b, a rocky, dense world that has been given the name Tahay, circling the M1 star that is now named Añañuca (Table 5.3). This star and its lava planet have been named after Chilean wildflowers because the flower Añañuca is red like the star and the flower Tahay only blooms for about 7–8 h a year, about the same as the planet's orbital period [30]. Given that any terrestrial vegetation on the surface of Tahay

Table 5.3 Planets of Añañuca (GJ 367)

Planet	Mass (Earth = 1)	Orbital period (days)	Distance from star (Earth distance from Sun = 1)	Equilibrium temperature (degrees K) (Earth = 255 K)	Type
GJ 367 b (Tahay)	0.63	0.32	0.0069	1262–1380	Very hot rocky
GJ 367 c	>4.1	11.53	0.076 (estimated)	380–416 (estimated)	Hot super-Earth
GJ 367 d	>6.0	34	0.16 (estimated)	262–287 (estimated)	Warm or hot super-Earth or sub-Neptune

Fig. 5.3 Artist's impression of the lava world 55 Cancri e (see Chap. 6). Reprinted from www.jpl.nasa.gov. © 2016, NASA/JPL-Caltech. Public domain

would be rapidly charred to a cinder, fields of wildflowers there are unlikely (Fig. 5.3).

We know both the radius and mass of Tahay, and the most recent measurements give its radius at about 70% of Earth's and its mass at about 63% of Earth's. Its density is therefore about 10 g per cubic centimeter, or considerably denser than Mercury, the densest solar system planet [31]. Tahay must have a hugely massive iron and nickel core as this would be the only way to reach such a high density. It is one of the class of so-called "super-Mercuries", rocky, exceptionally dense planets that are larger than Mercury but have similar compositions. Super-Mercuries do not seem to be very common, apparently because iron-rich planets can only be formed by an improbable series of high-energy impacts [32]. These collisions would be needed to blast away the mantle, the lighter, mostly solid but sometimes partly melted layer between the crust and the planet's core. The removal of the mantle would leave behind the denser iron core, and its exterior would then solidify and become the new surface of the planet.

Table 5.3 shows that the equilibrium temperature of Tahay is about 1300 K. This is about the minimum temperature for the maintenance of a magma ocean [33], but the planet is likely to be tidally locked, so the temperature at the sub-stellar point on the permanent day side of the planet may be several hundred degrees warmer than that. With a temperature approaching the melting point of iron, it is actually rather warmer than typical terrestrial lava. The day side temperature of the planet is high enough to remove any primordial hydrogen or helium atmosphere [34]. It is not yet known whether it has a substantial secondary atmosphere or not. Tahay is old, about 7 billion years, and unless its atmosphere were continually replenished by volcanoes, its secondary atmosphere would have been stripped a long time ago. If the atmosphere is thin or non-existent, the temperature on the night side of Tahay would be considerably lower than on the day side. How much lower is hard to say, but it would certainly be cool enough on the night side for lava to solidify. If the atmosphere is very thin, it might even be cold enough on the night side for there to be some sub-surface ice, as in the polar regions of Mercury. It would be a strange world: a magma sea and perpetual daylight on one side, permafrost and endless starlight on the other.

On Tahay, lava might be created by other processes than just simple melting of the surface by the intense heat of the star. A planet so close to its star would likely be strongly tidally heated if the orbit is at all eccentric. The latest value for the eccentricity of Tahay is substantially larger than zero, so intense tidal heating is likely [35, 36]. This would cause considerable volcanism, which would provide a lava source to replenish the continually evaporating lava seas, as well as a possible source for an atmosphere. Lava worlds would likely have low albedos, meaning that their surfaces would be really dark, although there might be some exceptions for worlds with less common surface compositions [37]. Lava worlds are conventionally defined to be smaller than 1.6 times the radius of Earth, as worlds bigger than that are more likely to retain a substantial atmosphere and would be more Neptune-like rather than rocky [38].

Tahay is hot, but it is not the hottest lava world known. Much hotter is Kepler-78 b, in the Vega sector about 125 parsecs away. This lava planet has an equilibrium temperature of more than 2000 K and a likely peak day side temperature much higher than that [39]. This temperature may approach or surpass the boiling point of iron, so the planet should have at least a thin planetary atmosphere comprising silicate and perhaps iron vapor. This type of atmosphere might be thick enough to transfer some heat to the night side, although estimates of the magnitude of this effect are at present very uncertain because there is little observational data to validate them. There would also be a magma ocean circulation, like the currents in Earth's ocean. Unlike the effect of ocean currents on Earth's climate, though, on very hot worlds this circulation does not really affect surface temperature, as it is utterly swamped by incoming stellar radiation [40].

Some estimates have been made of the geographical extent of magma oceans on lava worlds. In 2016, detailed modelling by Edwin Kite of the University of Chicago and his collaborators suggested that Kepler-78 b has a large magma ocean with an angular radius of about 100° [41]. This would cover the entire day side and even extend a little bit past the sunset line (or "terminator") into the night side. Using the same calculations, Tahay, being a lot cooler, would have a smaller but still substantial magma ocean several thousand kilometres across but not extending across the entire day side [33]. The surface elsewhere on the day side of Tahay would be solid, and the frozen lava seas might look like the lunar maria, like the flat Sea of Tranquility where Apollo 11 landed. They might also be seas of glass, as glasses are typically produced when lava solidifies rapidly (obsidian is an example) [37]. Magma ocean depths would be perhaps tens of meters on both planets, and on Kepler-78 b, an atmospheric pressure of up to 100 hPa is suggested, or about one tenth of the pressure at the surface of the Earth. At the sub-stellar point, gases would evaporate from the hot surface, then rise and travel towards the night side, where they would condense. The resulting rain could be silica or it could be iron. As I said in an article some time ago, this would bring new meaning to the term "heavy precipitation" [42].

The idea that the atmosphere of Kepler-78 b might be thick enough to be capable of transporting heat from the day side to the night side was given a boost by recent observations. Data was taken during the transit of the planet in front of its star and during the eclipse of the planet when it travels behind its star. During the eclipse, the infrared radiation that the planet emits is cut off, so the difference between the total amount of emission when the planet is hidden and when it is not a measure of the emission coming from the planet. During eclipse, the planet's day side is facing towards Earth, so this method gives an estimate of the day side temperature. When the planet transits in front of the star, its night side is facing towards Earth, so similar calculations can extract night side temperature as well. When this is done for Kepler-78 b, the night side temperature is estimated at about 2700 K [43]. This is much warmer than would be expected if there were no atmosphere, so the likely explanation is that Kepler-78 b still has an atmosphere, perhaps dominated by carbon dioxide, that is capable of transporting a large amount of heat from the day side to the night side. Compared to Tahay, Kepler-78 b is young, so its chances of retaining such an atmosphere are higher. This very high night side temperature also means that the magma ocean on Kepler-78 b is probably global, unlike its counterpart on Tahay.

Very young planets are also predicted to have magma oceans as a result of large impacts during their formation. Earth was like this a very long time ago, shortly after it formed. The enormous energy released by large impacts as they smashed into the proto-Earth provided a source of strong heating [33]. Later, a magma ocean also formed after the giant impact that created the Moon.

Actually, Earth still has a magma ocean: it is called the outer core and it lies about 3000 km below Earth's surface, below the thin solid crust and the much thicker semi-solid mantle [44].

But some lava planets are too hot for their own good. There are some that are literally evaporating into space. An example is Kepler-1520 b, circling a K star about 620 parsecs distant in the constellation Cygnus. This small, sub-rocky lava world probably has less than 400 million years to live [45]. The problem for this planet is that as it loses mass by evaporation, its gravity becomes less, making it easier to lose even more mass, and so on in a vicious cycle. Note, though, that planets as large as Earth have stronger gravity and so lose mass slower by this process, and thus are more protected from evaporation than small worlds. But not entirely. When the Sun becomes a red giant some 7.6 billion years from now, the Earth will become a lava world all over again, just before it is engulfed, and then likely becomes one of those worlds that evaporates [46].

*   *   *

Further Afield

Just outside the 10 parsec limit is another of Luyten's stars with a high proper motion, L 98-59. Located in Volans, the flying fish, a sparsely starred constellation located near the south celestial pole, this M3 star has four or five planets. Table 5.4 shows that this is a system with a large number of planets close to its star, different from the arrangement of planets in our solar system—but that is not unusual, because our solar system has an atypical arrangement of planets compared to most stellar systems [47]. The four innermost planets are located in or close to what has become known as the Venus Zone, or VZ, the region of a stellar system where a planet receives a stellar radiation amount similar to or higher than Venus [48]. The outer boundary of this zone is where the incoming radiation is high enough to cause a runaway greenhouse (Chap. 1). The inner boundary of the VZ occurs where the amount of incoming radiation is so high that the atmosphere starts to erode, at about 25 times Earth's incoming radiation. There are probably lots of VZ planets out there.

For the three innermost planets of the L98-59 system, we know their radii, masses and densities. Planet b has a density about 65% of that of Earth, so it is definitely a mostly rocky world. Attempts to characterize its atmosphere

Table 5.4 Planets of L 98-59

Planet	Mass (Earth = 1)	Orbital period (days)	Distance from star (Earth distance from Sun = 1)	Equilibrium temperature (degrees K) (Earth = 255 K)	Type
L 98-59 b	0.4	2.25	0.022	559–612	Hot rocky
L 98-59 c	2.2	3.69	0.03	479–524	Hot super-Earth
L 98-59 d	1.9	7.45	0.049	375–410	Hot super-Earth or Earth-size
L 98-59 e	>3.1	12.8	0.07	314–343	Hot super-Earth or sub-Neptune
L 98-59 f (unconfirmed)	>2.5	23.2	0.10	262–287	Warm or hot super-Earth

have so far been inconclusive [49]. Its spectrum is more consistent with a cloud-free, carbon-dioxide dominated atmosphere than a clear steam atmosphere. A cloudy steam atmosphere remains a possibility, and a small hydrogen/helium atmosphere is also not excluded at present. Another intriguing possibility is an oxygen atmosphere produced by photodissociation. Also possible is next to no atmosphere, like Mercury. This ambiguity may be resolved in the near future, as the system is nearby and thus easily observed by the JWST [50]. Watch this space.

Planet c has a density about 85% of Earth's, while the density of planet d is considerably less, at only 55% of Earth's. This is too low for the planet to be purely rocky and indicates either that it contains lots of water or has a thick atmosphere. At only 1.9 Earth masses, this planet is small to have the thick atmosphere of a sub-Neptune. Perhaps the waterworld scenario is more likely, implying a planet with a large, hot ocean. Results from the Hubble Space Telescope show that planets c and d do not have cloud-free hydrogen atmospheres [51]. Current realistic scenarios are that they have no atmosphere, a thin atmosphere that is not hydrogen or helium, or a hydrogen/helium atmosphere with a thick cloud layer.

There is one unconfirmed world in this system that might have more habitable temperatures. Planet f (if it exists) is a super-Earth with an equilibrium temperature only slightly higher than Earth's [52]. If this planet orbits in the same plane as the planets whose transits have already been measured, then planet f's minimum mass of 2.5 Earth masses would be about its actual mass. Without any information on its radius, though, its density cannot be quantified with any confidence, and so a wide range of compositions is currently possible.

Also in this sector just outside the 10 parsec limit is GJ 1132, yet another M star, this time in the constellation Vela, the sails. This collection of stars was originally part of the unwieldy super-constellation Argo. In 1752, recognizing that Argo was too large, the French astronomer Lacaille divided it into Vela, Carina (the keel) and Puppis (the poop deck). Argo was the ship that carried Jason and the Argonauts as they sailed the Aegean and the Black Sea in their quest to find and steal the Golden Fleece, an adventure that has spawned many adaptations. Among the most famous are the play *Medea* (431 B.C.E.) by Euripides, a dreadful tale of betrayal and revenge, and the classic sword-and-sandal film *Jason and the Argonauts* (1963) [53]. The film employs some very impressive stop-action animation and is fondly remembered by a generation of youngsters as a stalwart of weekend television.

Assessment of the actual surface conditions of the dense, mostly rocky Earth-size planet GJ 1132 b is problematic at this stage, as the relevant observations give conflicting information (Table 5.5). An atmosphere was announced as having been detected, but then later work did not confirm this [54]. Some more work then found the atmosphere again [55]. The eccentricity of its orbit is given as "< 0.22" in the paper announcing the discovery of planet c, which of course includes any value from 0.22 all the way down to zero. A value of 0.22 is big for a planet orbiting close to a star and would be certain to generate a large amount of tidal heating. In contrast, a value of zero would generate none. But for a world so close to its star, even an eccentricity of 0.01 would generate considerable volcanic activity [56]. This might result in a planet with a thin crust of less than a kilometer thick overlying a magma ocean, a so-called "eggshell world" [57, 58]. An eggshell world is defined as a planet with a crust so thin that it can bend, and this might lead to a planet with very little topography. A very thin crust might be caused by some combination of high surface temperature and internal heat flux, and in the case of GJ 1132 b, a high internal heat flux could be caused by tidal heating. A

Table 5.5 Planets of GJ 1132

Planet	Mass (Earth = 1)	Orbital period (days)	Distance from star (Earth distance from Sun = 1)	Equilibrium temperature (degrees K) (Earth = 255 K)	Type
GJ 1132 b	1.66	1.63	0.0154	527–576	Hot Earth-size
GJ 1132 c	>2.6	8.93	0.0476	300–328	Hot super-Earth
GJ 1132 d (unconfirmed)	>8.4	177	0.35	111–121	Cold sub-Neptune

mechanism that might stop the planet's orbit from being circularized by the relentless drag of tidal dissipation is gravitational resonance with planet c, in the same way that the eccentricity of Jupiter's moon Io is maintained by resonance with Jupiter's other moons. Even if it turns out that GJ 1132 b is not an eggshell world, there are likely to be numerous examples of this type of planet out there. Some rocky worlds that orbit close to their stars will have the eccentricity of their orbits maintained by resonances, and so will experience very strong tidal heating. The outgassing generated by tidal heating could lead to the continual regeneration of the atmospheres of eggshell planets, counteracting the stripping effect of the nearby star.

Eggshell planets, lava worlds—tidal heating can have big effects. With all of the discussion of lava planets in this chapter, it would be good if science fiction writers had written about one that was located within the bounds of the known galaxy, but there does not seem to be one. There is one in a galaxy far, far away, though. It would be remiss not to mention the fictional lava planet Mustafar, site of the epic lightsaber duel between Obi-Wan Kenobi and Darth Vader in Episode 3 of the *Star Wars* saga, *The Revenge of the Sith* [59]. Once a garden world, the planet has been turned into a tidally-heated hellscape by a series of traumatic events [60]. Despite this, it hosts a lively ecology adapted to its hostile environment, and a number of its resident life forms are highly aggressive. The darkghast, a multi-limbed, multi-mandibled carnivore, lives in caverns. The blistmok are reptiles that hunt in packs, tirelessly. There are giant eels that eat humans. Vegetation is sparse but hardy. The atmosphere is noxious and a breathing device is required, as fumes from the lava lakes cause hallucinations and delusions. Storms are frequent and deadly, and play havoc with communications and transport. Despite all of these issues, the planet is coveted for its mineral resources, and sure enough, some swashbuckling private companies have set up there. This hellish planet is also the location of Darth Vader's castle, where he could go to relax. It seems like the right place for him.

But back to the known galaxy. An unusual system in this sector is that of Gliese 229, about six parsecs away from Earth. It consists of an M star orbited by a T dwarf that was discovered in 1994, the first confirmed brown dwarf [61]. The T dwarf orbits the star at a mean distance of 28.9 a.u., but its very eccentric orbit means that it approaches as near as 4 a.u., or closer than Jupiter to the Sun. This close approach has not prevented planets from forming around the M star. One of these, planet c, is a warm super-Earth that receives about as much stellar radiation as Mars. Discovered only in 2020, very little is known about it, but it has become clear from the evidence provided by systems like that of Gliese 229 that planets can and do form in binary systems even when one of the stars makes a reasonably close approach to the other star.

Far out in this sector at about 95 parsecs distance is Canopus (Alpha Carina), also a former resident of the old constellation Argo. Canopus is the brightest star in the sector and the second brightest in the entire night sky, after Sirius. Named after the pilot of the ship of King Menelaus in the Trojan War, this A-type giant star is not known to have any real planets, but it hosts a most important fictional one. The planet Canopus III is Arrakis, the setting of Frank Herbert's classic novel *Dune* (1965) [62, 63]. Arrakis is a world that has no lakes or oceans and no precipitation but does have some subterranean water. Despite this, it has a breathable oxygen atmosphere. This is a strange combination, as the presence of oxygen would usually imply lots of green vegetation, and that would require rainfall.

This begs the question of whether a planet like Arrakis as described in the book could exist or not. The first requirement would be for the planet to orbit a star other than Canopus. Canopus is only about 25 million years old, and this is less than the time thought to be needed for the formation of a terrestrial planet [64]. Yet Canopus has already left the main sequence and become a giant star. Like all giant stars, Canopus has recently undergone a very substantial increase in luminosity that would have played havoc with the climate of any planet orbiting it, assuming that one exists. But let's instead assume that Arrakis circles a well-behaved, long-lived main sequence star. There are some limited circumstances under which a temperate, dry, lifeless planet could be furnished with an oxygen atmosphere. The recent simulations of planetary evolution by Joshua Krissansen-Totton of the University of Washington and collaborators (Chap. 4) examined the evolution of Earth-mass planets orbiting Sun-like stars [65]. Among the wide-ranging suite of possible planetary conditions generated were a smallish subset of worlds that were both dry and possessed abiotically-produced oxygen atmospheres [66].

The climate of Arrakis itself has been simulated [67]. Using a reasonably sophisticated three-dimensional climate model and making some brave assumptions about actual planetary conditions on Arrakis, some enterprising British-based meteorologists have found that while there are definite similarities between the simulated climate of the planet and the conditions described in *Dune*, there are also important differences. Unlike in the novel, simulated summer polar temperatures are much too high to permit the presence of ice caps. Also unlike in the novel, the simulated climate of Arrakis is not rainless, as the presence of some water vapor in the atmosphere is enough to spark off a small amount of summer precipitation in mountainous regions away from the equator. Finally, some regions of the planet have simulated climates that are too extreme for human habitation, with summers that are deadly hot and winters that are Siberian. I compare the simulated climate to the book climate in more detail in a recent article [66]. But the upshot is that a very dry world with a breathable oxygen atmosphere could exist, a world like Arrakis.

Sirius Sector Summary

This sector has the two brightest stars in the sky, a star that made a recent close approach to our solar system, a lava world, some planets in the Venus zone, a possible eggshell world, and the home star of Arrakis. But the sector is dominated by the old constellation Argo, still sailing across the sky. It was placed there after it accidentally dropped a wooden beam on its captain, Jason, killing him instantly. Argo has its place among the stars, but Jason does not. He betrayed his lover Medea to become engaged to a young princess, for political reasons, so the gods withdrew their favor from him. There are no constellations for traitors.

References

1. Ridpath I (2018) Star tales. Lutterworth, Cambridge, p 62
2. Wikipedia (2023) Sirius. Accessed 5 Oct 2023
3. Wikipedia (2023) Helical rising. Accessed 5 Oct 2023
4. Lucian, Hickes F (trans) (1902) Lucian's true history. Project Gutenberg ebook 45858. https://www.gutenberg.org/files/45858/45858-h/45858-h.htm
5. Wikipedia (2023) Micromegas. Accessed 5 Oct 2023
6. V (1983) imdb.com. https://www.imdb.com/title/tt0085106/?ref_=nv_sr_srsg_0_tt_8_nm_0_q_v%25201983. Accessed 6 Oct 2023
7. Von Reier S (2007) Im Rhythmus der Sterne. Die Zeit, 9 Dec 2007. https://www.zeit.de/online/2007/50/stockhausen-nachruf/komplettansicht. Accessed 21 May 2024
8. Lucas M, Bottom M, Ruane G, Ragland S (2022) An imaging search for post-main-sequence planets of Sirius B. Astronom J 163(2):81
9. Kratter KM, Perets HB (2012) Star hoppers: planet instability and capture in evolving binary systems. Astrophys J 753(1):91
10. Vanderburg A, Rappaport SA, Xu S et al (2020) A giant planet candidate transiting a white dwarf. Nature 585(7825):363–367

11. O'Connor CE, Liu B, Lai D (2021) Enhanced Lidov–Kozai migration and the formation of the transiting giant planet WD 1856+ 534 b. Mon Not Roy Astronom Soc 501(1):507–514

12. Luhman KL (2013) Discovery of a binary brown dwarf at 2 pc from the Sun. Astrophys J Lett 767(1):L1. https://doi.org/10.1088/2041-8205/767/1/L1

13. Garcia EV, Ammons SM, Salama M et al (2017) Individual, model-independent masses of the closest known brown dwarf binary to the Sun. Astrophys J 846(2):97. https://doi.org/10.3847/1538-4357/aa844f

14. Wikipedia (2023) Planckian locus. Accessed 7 Oct 2023

15. Pérez Garrido A, Lodieu N, Béjar VJS et al (2014) 2MASS J154043.42-510135.7: a new addition to the 5 pc population. Astronom Astrophys 567:A6. https://doi.org/10.1051/0004-6361/201423615

16. Cranmer SR (2021) Brown dwarfs are violet: a new calculation of human-eye colors of main-sequence stars and substellar objects. Res Not AAS 5(9):201

17. Ibid.

18. de la Fuente Marcos R, de la Fuente Marcos C (2018) An independent confirmation of the future flyby of Gliese 710 to the solar system using Gaia DR2. Res Not AAS 2(2):30. https://doi.org/10.3847/2515-5172/aac2d0

19. de la Fuente Marcos R, de la Fuente Marcos C (2022) An update on the future flyby of Gliese 710 to the solar system using Gaia DR3: flyby parameters reproduced, uncertainties reduced. Res Not AAS 6(6):136. https://doi.org/10.3847/2515-5172/ac7b95

20. Rampino MR, Caldeira K, Prokoph A (2019) What causes mass extinctions? Large asteroid/comet impacts, flood-basalt volcanism, and ocean anoxia—Correlations and cycles. In: Koeberl C, Bice DM (eds) 250 million years of Earth history in central Italy: celebrating 25 years of the geological observatory of Coldigioco. Geological Society of America, Boulder, CO

21. Siegel E (2018) What happens when stars pass through our solar system. Forbes, 31 Mar 2018. https://www.forbes.com/sites/startswithabang/2018/03/31/ask-ethan-what-happens-when-stars-pass-through-our-solar-system/?sh=671af4d3457f. Accessed 10 Oct 2023

22. When worlds collide (1951) imbd.com. https://www.imdb.com/title/tt0044207/?ref_=nv_sr_srsg_0_tt_8_nm_0_q_when%2520worlds%2520collide. Accessed 10 Oct 2023

23. Ridpath I (2018) Star tales. Lutterworth, Cambridge, p 108

24. Feng F, Butler RP, Shectman SA et al (2020) Search for nearby Earth analogs. II. Detection of five new planets, eight planet candidates, and confirmation of three planets around nine nearby M dwarfs. Astrophys J Suppl Ser 246(1):11. https://doi.org/10.3847/1538-4365/ab5e7c

25. Ibid.

26. Luque R, Pallé E, Kossakowski D et al (2019) Planetary system around the nearby M dwarf GJ 357 including a transiting, hot, Earth-sized planet optimal for atmospheric characterization. Astronom Astrophys 628:A39

27. Kaltenegger L, Madden J, Lin Z et al (2019) The habitability of GJ 357 d: possible climate and observability. Astrophys J Lett 883(2):L40

28. Permentier K, Vercammen S, Soetaert S, Schellemans C (2017) Carbon dioxide poisoning: a literature review of an often forgotten cause of intoxication in the emergency department. Int J Emerg Med 10(1):14

29. Chao KH, deGraffenried R, Lach M et al (2021) Lava worlds: From early Earth to exoplanets. Geochem 81(2):125735

30. IAU (2022) 2022 approved names. https://www.nameexoworlds.iau.org/2022approved-names. Accessed 12 Oct 2023

31. Goffo E, Gandolfi D, Egger JA et al (2023) Company for the ultra-high density, ultra-short period sub-Earth GJ 367 b: discovery of two additional low-mass planets at 11.5 and 34 days. Astrophys J Lett 955(1):L3. https://doi.org/10.48550/arXiv.2307.09181

32. Scora J, Valencia D, Morbidelli A, Jacobson S (2020) Chemical diversity of super-Earths as a consequence of formation. Mon Not Roy Astronom Soc 493(4):4910–4924

33. Chao KH, deGraffenried R, Lach M et al (2021) Lava worlds: From early Earth to exoplanets. Geochemistry 81(2):125735

34. Lam KW, Csizmadia S, Astudillo-Defru N et al (2021) GJ 367 b: A dense, ultrashort-period sub-Earth planet transiting a nearby red dwarf star. Science 374(6572):1271–1275

35. L'Observatoire de Paris (2023) GJ 367 b. In: Encyclopaedia of exoplanetary systems. http://exoplanet.eu/catalog/. Accessed 13 Oct 2023

36. McIntyre SRN (2022) Tidally driven tectonic activity as a parameter in exoplanet habitability. Astronom Astrophys 662:A15

37. Essack Z, Seager S, Pajusalu M (2020) Low-albedo surfaces of lava worlds. Astrophys J 898(2):160

38. Ibid.

39. Sanchis-Ojeda R, Rappaport S et al (2013) Transits and occultations of an Earth-sized planet in an 8.5 hr orbit. Astrophys J 774(1):54. https://doi.org/10.1088/0004-637X/774/1/54

40. Kite ES, Fegley B Jr, Schaefer L, Gaidos E (2016) Atmosphere-interior exchange on hot, rocky exoplanets. Astrophys J 828(2):80

41. Ibid.

42. Walsh K (2011) So long, Proxima Centauri. In: Analog Science Fiction and Fact, July/August 2011

43. Singh V, Bonomo AS, Scandariato G et al (2022) Probing Kepler's hottest small planets via homogeneous search and analysis of optical secondary eclipses and phase variations. Astronom Astrophys 658:A132

44. Young CJ, Lay T (1987) The core-mantle boundary. Ann Rev Earth Planet Sci 15(1):25–46. https://doi.org/10.1146/annurev.ea.15.050187.000325

45. Perez-Becker D, Chiang E (2013) Catastrophic evaporation of rocky planets. Mon Not Roy Astronom Soc 433(3):2294–2309. https://doi.org/10.1093/mnras/stt895

46. Williams M (2016) Will Earth survive when the Sun becomes a red giant? phys. org. https://phys.org/news/2016-05-earth-survive-sun-red-giant.html. Accessed 14 Oct 2023

47. Raymond SN, Izidoro A, Morbidelli A (2020) Solar system formation in the context of extra-solar planets. In: Meadows V, Arney G, Schmidt B, Des Marais DJ (eds) Planetary astrobiology. University of Arizona Press, Tucson, pp 287–324

48. Kane SR, Kopparapu RK, Domagal-Goldman SD (2014) On the frequency of potential Venus analogs from Kepler data. Astrophys J Lett 794(1):L5

49. Damiano M, Hu R, Barclay T et al (2022) A transmission spectrum of the sub-Earth planet L98-59 b in 1.1–1.7 μm. Astronom J 164(5):225

50. Pidhorodetska D, Moran SE, Schwieterman EW (2021) L 98-59: a benchmark system of small planets for future atmospheric characterization. Astronom J 162(4):169

51. Zhou L, Ma B, Wang YH, Zhu YN (2023) Hubble WFC3 spectroscopy of the terrestrial planets L 98–59 c and d: no evidence for a clear hydrogen dominated primary atmosphere. Res Astronom Astrophys 23(2):025011

52. Demangeon ODS, Zapatero Osorio MR, Alibert Y et al (2021) A warm terrestrial planet with half the mass of Venus transiting a nearby star. Astronom Astrophys 653:38. https://doi.org/10.1051/0004-6361/202140728

53. Jason and the Argonauts (1963) imdb.com. https://www.imdb.com/title/tt0057197/. Accessed 14 Oct 2023

54. Libby-Roberts JE, Berta-Thompson ZK, Diamond-Lowe H et al (2022) The featureless HST/WFC3 transmission spectrum of the rocky exoplanet GJ 1132b: no evidence for a cloud-free primordial atmosphere and constraints on starspot contamination. Astronom J 164(2):59

55. Swain MR, Estrela R, Roudier GM et al (2021) Detection of an atmosphere on a rocky exoplanet. Astronom J 161(5):213

56. Ibid.

57. Byrne PK, Foley BJ, Violay ME et al (2021) The effects of planetary and stellar parameters on brittle lithospheric thickness. J Geophys Res Planets 126(11):e2021JE006952

58. Smith A (2021) 'Lava egg' planet shocks Nasa scientists by creating 'second atmosphere'. The Independent, 12 Mar 2021. https://www.independent.co.uk/space/lava-egg-planet-second-atmosphere-nasa-b1816323.html. Accessed 18 Oct 2023

59. Star wars: Episode III—Revenge of the Sith (2005) imdb.com. https://www.imdb.com/title/tt0121766/?ref_=nv_sr_srsg_2_tt_1_nm_5_q_the%2520revenge%2520of%2520the%2520sith. Accessed 23 Oct 2023

60. Wookiepedia: the Star Wars wiki (2023) Mustafar. https://starwars.fandom.com/wiki/Mustafar. Accessed 23 Oct 2023

61. Oppenheimer BR (2014) Companions of stars: from other stars to brown dwarfs to planets and the discovery of the first methane brown dwarf. In: Joergens V (ed) 50 years of brown dwarfs—from prediction to discovery to forefront of research, Astrophysics and space science library, vol 401. Springer, Berlin, pp 81–111

62. Herbert F (1965) Dune. Chilton, Boston

63. Dune (2021) imdb.com. https://www.imdb.com/title/tt1160419/?ref_=nv_sr_srsg_3_tt_8_nm_0_q_dune. Accessed 23 Oct 2023

64. Lin DNC (2008) The genesis of planets. Sci Am 298(5):50–59

65. Krissansen-Totton J, Fortney JJ, Nimmo F, Wogan N (2021) Oxygen false positives on habitable zone planets around Sun-like stars. AGU Adv 2(2):e2020AV000294

66. Walsh K (2023) Dune and superdune. Analog Science Fiction and Fact, November/December 2023

67. Farnsworth A, Farnsworth M, Steinig S (2021) Dune: we simulated the desert planet of Arrakis to see if humans could survive there. https://theconversation.com/dune-we-simulated-the-desert-planet-of-arrakis-to-see-if-humans-could-survive-there-170181. See also https://climatearchive.org/dune#. Accessed 24 Oct 2023

6

Super-Earths and Superflares

The Procyon Sector: 6–12 h, 0 to +90°

Procyon and Sirius have a lot in common. Both are bright, nearby stars. Both head up constellations that are associated with canines: Sirius is Alpha Canis Majoris (Big Dog) and Procyon (Fig. 6.1) is Alpha Canis Minoris (Little Dog). Both are stellar systems that contain one main sequence star and one white dwarf. Both have been mentioned many times in science fiction, and both appear in some of the earliest literature that we know [1]. Procyon A is a main sequence star F star heading towards red giant status. At about 1.9 billion years old, Procyon A is older than Sirius A and is also less luminous. The white dwarf Procyon B is considerably dimmer, less massive but paradoxically slightly bigger than its cousin Sirius B. Procyon B was once an A star but left the main sequence about a billion years ago [2]. The two stars in the Procyon system orbit each other at a distance varying from as close as about 9 a.u., or about the distance from the Sun to Saturn, to as distant as 21 a.u., or a little further than the distance from the Sun to Uranus.

The name "Procyon" comes from ancient Greek, meaning "before the dog", as Procyon precedes Sirius in the night sky as the Earth rotates them both into view. It appears in the lists of stars and constellations made by the Babylonians before 1000 B.C.E [3]. Procyon has been the topic of numerous myths, but one of the most evocative is told by the Inuit, who call the star Sikuliarsiujuittuq (no, I don't know how to pronounce it) [4]. This means "the one who never goes onto the newly formed sea-ice" and refers to a man who was so fat that he refused to go hunting on the sea ice. Instead, he stole food from his fellow hunters. Needless to say, he met a violent end. Procyon sometimes appears blood-red when it rises during the Arctic winter, serving as a reminder of what might happen in that uncompromising environment.

© The Author(s), under exclusive license to Springer Nature Switzerland AG 2024
K. J. E. Walsh, *Planets of the Known Galaxy*, Science and Fiction,
https://doi.org/10.1007/978-3-031-68218-6_6

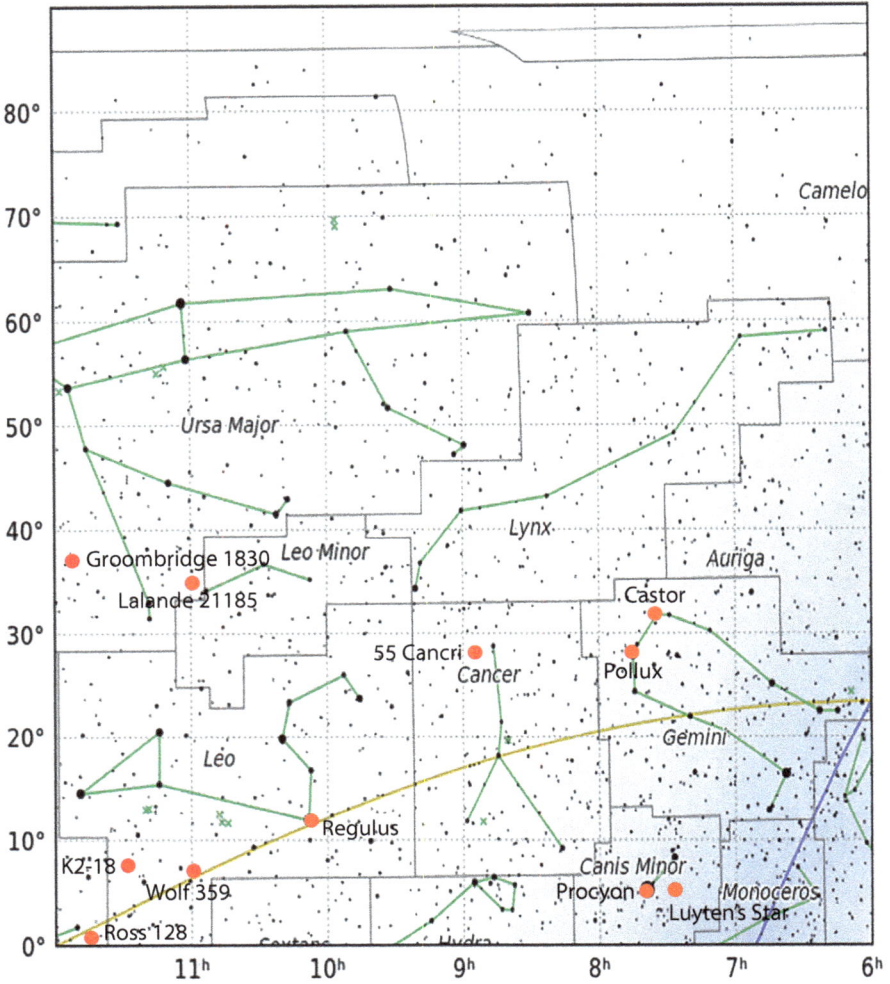

Fig. 6.1 Map of the Procyon sector. Blue shading indicates the "Milky Way", the most star-packed part of the galaxy, while the blue (darker) line is the galactic equator, an imaginary circle running through the Milky Way that slices the galaxy in half. The orange (lighter) line is the path of the Sun's location through the seasons (indicating the "zodiac" constellations). Larger circles indicate brighter stars, while locations mentioned in this chapter are indicated by red circles. Image produced using StarCharter (https://github.com/dcf21/star-charter, accessed on the 2nd of July 2024. © 2020, Dominic Ford) software under the GNU general public license

More recently, the Procyon system became the home of a planet in Larry Niven's Known Space universe named "We Made It", so-called because its colonists crash-landed there. The Crashlanders, as they came to be known, inhabit a world that is not entirely hospitable. It is tilted on its side like

Uranus, leading to bizarre patterns of seasonal and daily weather variations, including winds so ferocious that for part of the year the inhabitants have to live underground [5].

Only about 0.3 parsecs from Procyon is Luyten's Star, an ancient M star named after U.S.-based astronomer Jacob Luyten, who in 1935 first discovered that it had a high proper motion. Luyten made a number of significant advances in the measurements of nearby stars. According to those who knew him, he was a real character, a gruff, irascible man who often mocked his colleagues and as a result alienated many of them [6]. He also seemed to think that certain stars were "his" stars, and that no one else had a right to investigate them. Since Luyten's star has been named after him, it really is his star, and has at least two and possibly four planets, including one at the inner edge of the conservative habitable zone. These planets do not transit and so we only have radial velocity measurements, and usually this would only give us a minimum mass, not the actual mass. But analysis of the gravitational stability of various combinations of orbits can significantly constrain the allowed masses of the planets [7]. For the case of four planets, very few orbital configurations are actually stable. The only ones that are stable are those where the orbits of the two larger, outer worlds have zero eccentricity and high, almost edge-on inclinations.

Knowing this, Francisco Pozuelos of the University of Liège in Belgium and his co-authors have used a model of planetary formation to investigate the types of planets that might be formed in this system. To do this, they use the MERCURY software, a long-established piece of code that simulates the movement of large objects around a stellar system and how they interact gravitationally, as well as how they collide and merge to form planets [8]. This code has been used to construct artificial stellar systems to try to get a feel for the kinds of systems and planets that are likely to develop from the chaos of stellar system formation. The model starts with a large number of planetary "embryos", asteroid-like objects about some hundreds of kilometers wide or larger. An initial distribution and composition of the embryos is then specified, the simulation begins, embryos collide, and planets form. Planets acquire circular or eccentric orbits, are ejected from the system, and end up covered in water or with no water at all.

Naturally, the end result varies wildly from simulation to simulation, and depends significantly on the assumptions made in the model. Nevertheless, some conclusions can be made from the model results about the mechanisms that are likely to operate in specific stellar systems. For the planets orbiting Luyten's star, those simulations that produced planets resembling planets b

Table 6.1 Planets of Luyten's Star (GJ 273)

Planet	Mass (Earth = 1)	Orbital period (days)	Distance from star (Earth distance from Sun = 1)	Equilibrium temperature (degrees K) (Earth = 255 K)	Type
Luyten's Star c	1.2	4.72	0.036	424–464	Hot Earth-size
Luyten's Star b	2.9	18.64	0.091	267–292	Warm or hot super-Earth
Luyten's Star d	10.9	414	0.712	95–104	Cold sub-Neptune or sub-Jovian
Luyten's Star e	9.4	542	0.849	87–96	Cold sub-Neptune or sub-Jovian

and c indicate that planet b is probably a water world, while planet c is a world with low water content (Table 6.1).

Massive, water-rich super-Earths like planet b are likely to be common in the known galaxy. A deep ocean has the huge advantage of making seasonal variations less than they would otherwise be if the planet were mostly land, thus improving prospects for habitability. This is because water heats up much slower than land does, just as the water in a saucepan heats up much slower than the saucepan itself. This would still be the case even if the planet were tilted on its side like Uranus, with a tilt of about 90° instead of Earth's 23.5°. In a world like that, the north pole would experience 6 months of daylight followed by 6 months of light, just like it does on Earth. But on a world tilted on its side, as mid-summer approached, at the poles the Sun would get higher and higher in the sky. During mid-summer at the poles, the Sun would be almost directly overhead, like in Earth's tropics and unlike in Earth's Arctic, where the mid-summer polar summer sun is low and near the horizon. As a result, on this sideways-spinning world, the poles would be much more strongly heated during summer than on Earth.

Climate model simulations of a waterworld Earth tilted on its side were published in 2014 by David Ferreira of M.I.T. and collaborators [9]. These were the first of their kind to use what is known as a coupled ocean-atmosphere general circulation model, one that is able to simulate atmospheric winds, oceanic currents and the interactions between them. They found that the coldest region of the sideways planet is the equator, but surface temperatures there still remain above freezing. In the winter part of the planet, the part facing away from the Sun, temperatures are also well above freezing, with typical polar temperatures about 10 °C (50 °F). There are also very weak winds in this region because of the weak temperature variations from place to place caused by the very large area of zero incoming stellar radiation over the winter pole. In summer, mean temperatures there are above 30 °C (86 °F) but with a greater day-to-day variation than in winter. This is hot by terrestrial standards and on an aquaplanet (a planet with no land) it would also be humid. These are still eminently habitable conditions, however.

But there is a catch. Climate simulations have given us an idea of the various climates that can be created on a planet by changes in axial tilt. Oddly enough, simulations have shown that maximum habitability appears to be at about 55° tilt, not at Earth's current tilt of about 23° [10]. At a tilt of 55°, while seasonal variations of temperature are large at many locations, a planet can have less incoming radiation and still not completely freeze over, as the polar regions become too warm in the summer for permanent ice caps to form [11]. This makes it difficult for the planet to turn into a snowball. Also, at tilts greater than this, the amount of incoming solar radiation is actually greater at the poles than at the equator. In that situation, the permanent ice can be at the equator, not the poles. This forms an "ice belt" around the equator rather than an ice cap at the poles. Ice belts may in fact be more common on habitable planets around G and K stars than ice caps [12]. This is because many habitable planets will undergo periodic oscillations in their axial tilt over thousands or millions of years, just as Mars has thought to have undergone, with many planets ending up with a period of high axial tilt at some point. Earth does not undergo such dramatic oscillations because the Moon stabilizes these variations, but many habitable worlds will not have a large satellite like our Moon [13].

Not only does the tilt of the axis affect habitability, so does the rotation rate. Simulations show that slowly rotating worlds are cooler than fast rotating ones. This is because intense heating of the surface generates convection, the upward movement of parcels of moist, warm air. As the parcels rise, they cool, eventually condensing into cloud, creating persistent showers and thunderstorms. The sub-stellar point on a slowly rotating world does not move very fast, so there is a lot of time for cloud to form, and eventually the sky in that region becomes very cloudy, along with much of the surrounding day side [14]. At present, information on actual axial tilt and rotation rates for Earth-sized exoplanets is lacking, apart from the rotation rates of worlds that are very likely to be tidally locked. So far, rotation rates have only been measured for some large Jovian worlds (e.g. Beta Pictoris b, with a day length of only about 8 h [15]). But it is possible in principle to measure rotation rates of Earth-sized exoplanets, when larger and better telescopes become available in the future [16]. For fast-rotating planets, direct imaging techniques have the potential to detect the rotation of albedo features like land and clouds and the resulting changes in reflectivity, and thereby determine the rotation period.

Luyten's star is the location of another of Larry Niven's Known Space planets. The inhabited part of the planet Down has a climate similar to the desert regions of southern California. Farmers need UV lamps to grow crops in the red light of its star [17]. It is inhabited by Grogs, largely inanimate creatures that look a little like Cousin It from the Addams Family. As usual in science fiction, the Grogs are more mysterious than they seem at first. More recently, a planet around Luyten's Star is part of the universe of the Starfield video game, released in 2023 [18]. This enormous game contains about 1000 planets, including Luyten's Rock, a temperate Earth-sized world with a carbon dioxide atmosphere. Its traits include a crystalline crust, where magma wells up through mineral water to form large crystals [19]. Unfortunately, none of the real planets of Luyten's Star fit this description, but maybe there is room between planets b and d for an undiscovered world.

Luyten's Star b *How similar to Earth: Larger, and likely a water world. Also, likely hotter.*
Plausible planet: Warm to hot water world.
Best case scenario: Mild temperatures in many parts of the planet.
Worst case scenario: Thick atmosphere, unbearably hot everywhere.
Cultural connection: The planets Down and Luyten's Rock.

*   *   *

Superflares are large, energetic flares that are emitted by solar-type stars. Even the Sun has small superflares, very occasionally. In 1859, a large solar flare, now known as the Carrington event after the amateur astronomer who first noticed it, played havoc with telegraph communications. It caused sparks to shoot from telegraph machines and gave their operators electric shocks [20]. The northern lights flared up and were seen as far south as Hawaii. In some locations, the auroras were so bright that they were mistaken for dawn [21]. Some telegraph operators found that they could run their telegraph lines without batteries, just by using the current induced in the line by the auroras. The damage to the nascent telecommunications infrastructure of the nineteenth century was significant, but if a repeat of the Carrington event were to occur today, it would wreck the Internet. Estimates of the potential damage run into the trillions of dollars [22].

Compared to superflares on some other stars, the Carrington event was rather tame. Take the G subdwarf Groombridge 1830, in this sector about 9 parsecs from Earth, in the constellation Ursa Major. It is called a subdwarf because it is much less luminous for its mass than a main sequence star, and the culprit is its low abundance of elements heavier than hydrogen or helium. Astronomers call any element that is not hydrogen or helium a "metal". To paraphrase Isaac Asimov, who once said that the solar system consisted of one star, four planets, plus some debris [23], the elements hydrogen and helium are so dominant in the universe as a whole that the universe could be considered to consist of hydrogen and helium, plus debris. In 1939, astronomers at the Allegheny Observatory in Pittsburgh recorded a flare on Groombridge 1830 that was about 100 times more energetic than the Carrington event [24, 25]. This was not even the largest superflare so far observed on a Sun-like star. In 1899, the Sun-like star S Fornacis, about 120 parsecs away in the southern hemisphere constellation Fornax, was observed to emit a flare with an energy calculated to be more than 100,000 times as energetic as the Carrington event. This superflare caused the star to become temporarily more than ten times brighter than it usually is.

Imagine what might happen to life on Earth if for a couple of hours the Sun became ten times brighter. On the day side over land, a quick calculation shows that surface ground temperatures would rapidly soar by more than one hundred degrees Celsius. Plants would shrivel. People and animals who did not take shelter underground would probably die. The night side of the planet would not be immune either, as there would be catastrophic damage to electronic infrastructure. Such an event could plunge our world into a new Dark Age. On the other hand, if the flare occurred when it was noon over the middle of the Pacific Ocean, simulations indicate that sea surface temperatures there would not be strongly affected, due to the very large heat capacity of the ocean and the short duration of the flare [26]. Further possible impacts of a superflare on life on Earth were examined in 2017 by scientists Manasvi Lingam and Abraham Loeb, when Lingam was at Harvard [27]. The effects of a flare about 20,000 times more energetic than the Carrington event might include destruction of the ozone layer, very significant rises in surface temperatures, and unknown but probably highly detrimental effects on a wide range of ecosystems. A mass extinction event would not be out of the question.

It is not certain whether such large flares have actually occurred on the Sun, however. There is nothing in the historical or fossil record that suggests that Earth has ever suffered from a flare of this magnitude. To get a handle on the likely occurrence of superflares on our Sun, examination of a large number of solar-type stars is needed. This is exactly what has been done by Soshi Okamoto of the University of Kyoto and his team [25], who examined the Kepler data as well as the data from the European Space Agency Gaia space telescope. Like the JWST, Gaia resides at one of the so-called Lagrangian points of Earth's orbit. Named after the very eminent eighteenth and nineteenth century scientist Joseph-Louis Lagrange, these locations are where the gravitational attraction of the Sun and the Earth, combined with the centrifugal force created by the orbit of the spacecraft about the Sun, cancel each other out. Objects at these Lagrange points therefore do not move around much relative to the Earth, and so the amount of on-board fuel required for orbit corrections is kept to a minimum.

There are several of these Lagrange points, and the Gaia telescope is located at the nearest one, the L2 point about 1.5 million kilometers further away from the Sun than Earth [28]. From this vantage point, the Gaia telescope can measure the positions of stars with great accuracy. In principle, this method could be used to detect planets because as they orbit their star, their gravitational attraction pulls the star from side-to-side. Unfortunately, the movements are tiny, and until recently this astrometry technique was not competitive with radial velocity measurements (see Chap. 1). The Gaia space telescope now performs these measurements with unprecedented precision, about 100 times better than its revolutionary predecessor, the Hipparcos satellite [29]. With the advent of Gaia, astrometry becomes a more routine technique for finding exoplanets. Unlike radial velocity measurements, astrometry works well for large planets in distant orbits, because a large planet in a distant orbit gives a larger wobble than the same planet in a close orbit. Astrometry also provides an accurate estimate of the mass of the planet, not just the minimum mass, a limitation of the radial velocity method. Dozens of candidate planets have been found by Gaia using astrometry. At least one candidate planet is in this sector and is relatively close, a super-Jovian planet 15.7 parsecs away orbiting the white dwarf GD 140, discovered in 2022 [30].

As Gaia is able to measure the positions of stars in the sky with great accuracy, it can also be used to calculate better estimates of parallaxes and thereby stellar distances. In their study on superflares, Okamoto and colleagues used these improved distance estimates from Gaia to get better measurements of the size of stars, to focus on ones that were similar to our Sun. This analysis, combined with brightness measurements from thousands of Kepler stars, enabled them to detect a large number of superflares, over 2000 of them from 266 solar-type stars. A more restricted subset of Sun-like stars was then defined by using only a subset of all G stars centered on classification G2 (the Sun's classification), and assuming a rotation rate and age similar to the Sun. For these stars, superflares with energies roughly 100 times that of the Carrington event were observed. If the observed flaring rate of these stars is similar to that of our Sun, statistical analysis implies that a flare with about 20 times the energy of the Carrington event should occur on our Sun about every 6000 years or so.

The idea that our nurturing Sun could deal civilization on Earth a crushing blow is discouraging. In fact, it seems almost unfair. But few things about the known galaxy are fair.

*   *   *

As the seventh-closest stellar system to Earth, Lalande 21185 has received plenty of attention from science fiction writers. It is the location of Hal Clement's novel *Star Light* (1971) [31], a sequel to *Mission of Gravity* (Chap. 3), also set on a world of extremes. The large planet Dhrawn has a very slow rotation, a long, eccentric orbit around its star, lots of internal heat, and a surface gravity 40 times that of Earth. Who could possibly explore such a world? The flat, centipede-like intelligent natives of high-gravity Mesklin from *Mission of Gravity*, naturally. As on Mesklin, there are some strange goings-on in the atmosphere of Dhrawn. The ammonia in the atmosphere combines with water to form a mixture that freezes at a much lower tempera-ture than pure water. Ammonia fog interacts with surface water snow and melts it, making surface transportation difficult. Weather forecasting on this world of constantly changing melting points is a difficult challenge, to say the least (even more so than on Earth…).

In Gregory Benford's highly detailed, dramatic and inventive hard science fiction novel *Across the Sea of Suns* (1984), two planets are detected in the Lalande 21185 system, a super-Jovian and a terrestrial world in the habitable zone [32]. Radio transmissions have been found coming from the habitable planet, so a spaceship is sent to investigate. The habitable planet, Isis, is tidally locked and orbits its star every 29 days. It is described as reddish on its sun-ward side and covered in glaciers on the night side. Along the terminator, icebergs carve into a pink sea that separates the ice from the land. The sub-stellar region is called the Eye, a bare, baked region of yellow sulphur dioxide plains and dunes, wind-swept by acid dust. The Eye stares at its star and is surrounded by an eye socket of mountains, snow-capped and cut by faults. Oxygen is in short supply in the atmosphere of Isis, and the local life exhausts itself trying to scavenge breathable air from the noxious gases. On arrival in

Table 6.2 Planets of Lalande 21185

Planet	Mass (Earth = 1)	Orbital period (days)	Distance from star (Earth distance from Sun = 1)	Equilibrium temperature (degrees K) (Earth = 255 K)	Type
Lalande 21185 b	>2.6	12.93	0.0788	349–381	Hot super-Earth
Lalande 21185 d	>3.9	215.7	0.5141	137–149	Cold sub-Neptune
Lalande 21185 c	>13.6	2946	2.94	57–62	Very cold sub-Jovian

this system, the Earthlings attempt to contact the local aliens, but this does not go well, and hints at a more deadly danger to come.

In reality, Lalande 21185 has two and possibly three planets (see Table 6.2). One notable feature of this system is that it hosts a sub-Jovian (planet c) on an orbit that is slow for planets of an M star, with a period of almost 8 years. This is not the longest orbit known around an M star of about this mass: the sub-Jovian Groombridge 34 c (Chap. 7) has an orbit of 7555 days and receives less incoming radiation from its primary than Neptune does from the Sun. More such planets with long orbits are likely to be discovered, and the recent improvements in astrometry should at least partially address this issue.

Like Barnard's Star, Lalande 21185 has been the subject of early claims of detected planets that did not pan out. As for Barnard's Star, the technique used was astrometry. George Gatewood of the University of Pittsburgh, one of the astronomers who debunked the previous claims of a planet around Barnard's Star, announced in 1996 that there were planets orbiting Lalande 21185 [33]. No planets with the parameters listed in that paper have been found since. Having said this, Gatewood has made numerous contributions to astronomy during his long career, becoming a recognized expert in the properties of nearby stars. No scientists are right all of the time (luckily, they are right are lot more often than they are wrong).

The smallish super-Earth Lalande 21185 b is hot by terrestrial standards, so it probably is not a prime target for astronomers trying to find out whether it might host life or not. That doesn't stop them speculating on whether techniques from left field might assist in this process. For instance, Thomas Beatty of the University of Arizona published a paper in 2022 examining whether city lights from an alien civilization might be detectable on the night side of nearby extrasolar planets [34]. Now before we start wondering about the quality and quantity of alien nightlife, the study concluded that the current level of illumination provided by Earth's city lights would not be detectable by the next generation of space-based telescopes if similar instruments were being used by aliens based in another solar system. Nevertheless, if the alien cities on some planets emitted more light than cities on Earth, that would be detectable. The easiest detection would be on the closest known planet, Proxima Centauri b, with light levels only needing to be about 10 times that of Earth. For Lalande 21185 b, light levels would need to be about 60 times brighter. Whether the nightlife would therefore also need to be 60 times better is not known.

Lalande 21185 b *How similar to Earth: Not very. Hot super-Earth, very likely too warm to be habitable.*
Plausible planet: Runaway greenhouse.
Best case scenario: Dry planet that might be more habitable on its night side.
Worst case scenario: Venus-like world.
Cultural connection: Isis and Dhrawn.

*   *   *

In the bottom left corner of Fig. 6.1 are two M stars that host planets. The first is another one of the Wolfs, Wolf 359. Located in the zodiacal constellation Leo at a distance of only 2.4 parsecs, Wolf 359 is the sixth-closest known stellar system (as of 2024). It appears briefly in one of the best Star Trek plot lines ever, two episodes of *Star Trek: Next Generation* later put together as the movie *The Best of Both Worlds* [35]. Captain Picard and his intrepid crew face a dire threat from the Borg, humanoids who assimilate the brains of their captives into their collective consciousness. When Picard himself is captured, the crew of the Enterprise face a terrible choice. In the novel *Across the Sea of Suns*, this system hosts an ancient, small world with a thin atmosphere and a declining biosphere. In reality, this star is relatively young and flares often. It hosts at least one and possibly two planets, or possibly no planets—it depends who you believe at this time [36]. Either way, not much is known about them.

Only a little more than a parsec from Wolf 359 is another system that plays a part in *Across the Sea of Suns*. In Benford's novel, the Ross 128 system has five gas giants and two Earth-sized worlds. Circling one of the gas giants is a moon similar to Ganymede named Pocks, a world that shares a disturbing connection to the disasters that befell the Earthlings at Lalande 21185. In reality, this very dim and inactive M star located in Virgo hosts Ross 128 b, an Earth-sized planet only slightly hotter than Earth. Ross 128 b has an equilibrium temperature of about 300 K, close to Earth's value of 255 K, but also warm enough for runaway greenhouse to occur, perhaps [37]. Modeling of the formation of this system suggests that this planet formed with a very large water fraction, due to its initial position beyond the snow line, followed by its inward migration [38]. Unfortunately, as Ross 128 b does not transit, we do not know its density accurately, and that would go a long way towards determining whether it has a high water mass fraction or not. Nevertheless, if we assume that it has a high water content, along with the same abundance of metals as its star, some modeling of the interior structure of the planet can be performed [39]. Like the Earth, Ross 128 b has a core, but one that is possibly larger than Earth's. Also like Earth, Ross 128 b has a mantle, but unlike Earth above the mantle is a layer of ice, then a very deep ocean. Unfortunately, a high water mass fraction means less iron, implying a weaker magnetic field. The slow rotation rate of the planet does not help, so stripping of its atmosphere is a definite possibility.

It goes without saying that planets with high water mass fractions could have deep oceans. In Chap. 2, we noted that a deep ocean would have an ice layer at the bottom, and this might wreck the carbonate-silicate cycle that acts to stabilize the planet's climate (Fig. 2.4). Nevertheless, recent work has suggested that the prospects for the long-term habitability of water worlds like Ross 128 b are better than previously thought. In 2018, Edwin Kite and Eric Ford of the University of Chicago and Penn State respectively performed some detailed calculations on whether water worlds would remain habitable over billions of years [40]. They made a generous definition of "potentially habitable" as a planet that has oceans cooler than 450 K (i.e. still very hot) and that last for more than 100 million years (geologically, not very long). Surprisingly, they found that the carbonate-silicate cycle does not really need to be present on water worlds for the climate to remain stable over geological time scales. With certain combinations of atmospheric parameters, water worlds can have a stable climate for more than 1 billion years. The reasons are rather technical but revolve around the interaction between changes in surface ocean temperatures and their effect on atmospheric carbon dioxide content, and how under certain circumstances this interaction can lead to a more stable climate. The mechanism goes like this. Let's say carbon dioxide started to increase in a water world atmosphere. Initially, this would increase the greenhouse effect and act to warm the planet. But when the carbon dioxide atmospheric surface pressure exceeds 1 bar, or about the total atmospheric pressure at the surface of the Earth, the scattering of incoming visible radiation by the carbon dioxide molecules becomes very significant. This ends up reflecting more visible radiation away from the planet than the resulting increased greenhouse effect, cooling the planet rather than warming it, thus stabilizing the climate. This study was performed for Sun-like stars, but the authors also discuss whether waterworlds around M stars might have similar habitability issues. The main potential problem, as we have already mentioned, is atmospheric stripping.

Then there is tidal heating. As Ross 128 b is a close-orbiting world with a very slightly eccentric orbit, it would likely experience tidal heating. A recent estimate suggests heat flow values on Ross 128 b are typical of very geologically active regions on Earth, or about 100 times Earth's average rate [41]. Now a word about these tidal heating estimates. As researchers themselves acknowledge, these are quite inaccurate at this time. Tidal heating rates go up as the square of the eccentricity, and many eccentricity values of extrasolar planets are not known precisely. So if the eccentricity is assumed to be 0.1 instead of the actual (but unknown) value of 0.01, the calculated tidal heating rate would be 100 times greater than it should be [42]. The calculated heating rates also rely on approximations regarding how the interior of planets might heat up as a result of tidal friction, and these processes are poorly understood. As a result, these estimates need to be taken with a grain of salt. Nevertheless, these calculations provide useful estimates that, even if they are wrong by a factor of more than 10, still can tell us something about the general planetary environment, particularly for worlds whose orbits and masses are known accurately.

Ross 128 b thus may be a volcanically active, hot, tidally locked water world with lots of undersea volcanism. All of this activity would be occurring at the bottom of a very deep ocean, under a layer of high-pressure ice formed by the immense water pressures at those depths. Volcanism beneath ice also occurs on Earth. Known as "glaciovolcanism", a recent example was the Icelandic volcano Eyjafjallajökull, which erupted under an Icelandic ice cap in 2010. A related phenomenon, cryovolcanism, is the eruption of water or other substances that would typically be frozen at the surface temperature of the world where it is occurring. This process occurs on several worlds in the outer part of our own solar system, including Saturn's moon Enceladus (Fig. 6.2). Thus it is likely to be common in the known galaxy. Recent work by NASA's Lynnae Quick and her collaborators has examined the prospects for cryovolcanism on a number of nearby planets that might be "cold ocean planets", or worlds with oceans that are largely or completely frozen [43]. Making a number of assumptions about the mostly unknown properties of these worlds, calculations are made about how thick their ice shells would be and therefore how amenable they would be to geyser-like eruptions of water. For Proxima Centauri b (Chiron's Waterworld), a very thin ice shell of less than 100 meters thick was calculated, suggesting that this might be a planet with considerable regions of unfrozen ocean. If there is some tidal heating, this world might have lots of geysers. They would be difficult to detect, however, as Chiron's Waterworld does not transit, making measurement of its

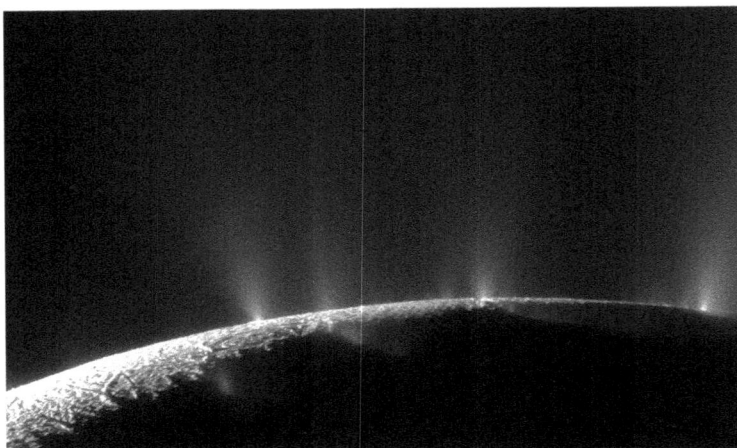

Fig. 6.2 Geysers on Saturn's moon Enceladus. eprinted from https://photojournal.jpl.
nasa.gov/. © 2010, NASA/JPL/Space Science Institute. Public domain

atmospheric spectrum more difficult. Arguably the most interesting of the
worlds that were examined in this study are the TRAPPIST-1 planets, but we
will talk about them in Chap. 9.

Another plausible scenario for Ross 128 b is runaway greenhouse, as it
receives more radiation than the greenhouse limit for wet planets, but not for
dry ones [44]. Or perhaps its atmosphere was stripped a long time ago. One
day soon, we will know.

Ross 128 b *How similar to Earth: Slightly hotter, likely wetter.*
Plausible planet: Hot, volcanically active water world.
Best case scenario: A tropical water world.
*Worst case scenario: Runaway greenhouse, or stripped atmosphere resulting in an
almost airless planet.*
Cultural connection: Pocks.

*   *   *

Further Afield

At 12.6 parsecs away in this sector, the 55 Cancri system has been particularly
well studied. There are two stars in it, the K star 55 Cancri A, otherwise
known as Copernicus, and about 1000 a.u. away the much smaller M star 55
Cancri B. The known planets in this system orbit Copernicus. Like many of

Table 6.3 Planets of 55 Cancri A (Copernicus)

Planet	Mass (Earth = 1)	Orbital period (days)	Distance from star (Earth distance from Sun = 1)	Equilibrium temperature (degrees K) (Earth = 255 K)	Type
55 Cancri e (Janssen)	8.6	0.74	0.01544	1790–1956	Very hot super-Earth or sub-Neptune
55 Cancri b (Galileo)	266	14.65	0.1134	660–722	Hot sub-Jovian
55 Cancri c (Brahe)	>56	44.4	0.2373	456–499	Hot sub-Jovian
55 Cancri f (Harriot)	>47	260.9	0.7733	253–276	Warm or hot sub-Jovian
55 Cancri d (Lipperhey)	>1217	5574	5.957	91–100	Cold super-Jovian

the newly discovered stellar systems, it is compact compared with our solar system, with three of its planets closer than Mercury is to our Sun. These three planets are also not small, with all of them considerably more massive than Earth (see Table 6.3). By far the best known is the innermost one, 55 Cancri e. It now goes by the name of Janssen, after the Dutch spectacle maker Zacharias Janssen (1585–c.1630), possibly one of the inventors of the telescope, but also likely a counterfeiter of coins [45].

A big world with a density slightly greater than Earth's, Janssen is a lava world (Fig. 5.3). That status is guaranteed by its very high equilibrium temperature, but also because Janssen is one of the few extrasolar planets whose surface temperature has been measured. Earlier results indicate that the sunward side of this tidally locked world experiences an eye-watering 2700 K, with some analyses suggesting temperatures in excess of 3000 K [46]. More recent results suggest a value of about 1800 K. This is a lot lower than the value calculated for zero albedo and no transport of heat to the night side, implying that there might be an atmosphere [47]. There is a consensus that the planet does not have a hydrogen or helium envelope, as this is ruled out by transit spectroscopy. The atmosphere might be a thin one of evaporated rocks, but if so, the non-detection of iron vapor would indicate a surface atmospheric pressure of less than 100 mb [46]. The most recent results suggest instead a thickish atmosphere (<100 bar) of carbon monoxide or nitrogen, with some carbon dioxide [47].

Another possibility is suggested by the composition of Janssen's host star. It has a higher fraction of carbon than our Sun, and this might be reflected in the composition of its planets. Thus it is not impossible that Janssen might be carbon rich and water poor [48]. A planet with lots of carbon would also inevitably have lots of graphite and diamond, as these are forms of carbon. It might have volcanoes that erupt diamonds and oceans made of liquid tar [49]. Perhaps sailors on those seas really would be called tars. Irrespective, the planet is a high-priority target for the JWST, so some of these questions are likely to be answered soon.

As for the other planets in the system, the hot Jovian called Galileo (55 Cancri b) may have hydrogen in its atmosphere. The next two planets outwards from the star are sub-Jovians, and one of them is in the conservative habitable zone [50]. The planet Harriot (55 Cancri f) is named after Thomas Harriot, an English astronomer who traveled with the 1585 expedition to North America funded by Sir Walter Raleigh, and who made written telescopic observations of the Moon some months before Galileo did [51]. Overall, the chance that the roughly Neptune-sized Harriot might host a habitable moon appears to be slim, but not zero. Formation of a habitable moon around Harriot by the same mechanism that formed Jupiter's larger moons, aggregation of smaller rock particles in the same way that our solar system planets were formed, seems unlikely, as it would lead to moons that are probably too small to have Earth-like habitability [52]. Larger moons could be established in orbit by the aftermath of a grazing collision (like Earth's Moon) or by capture (like Triton, the large moon of Neptune) [53].

Another obstacle is whether the orbit of an exomoon is stable over geologic time, due to tidal and gravitational effects (Chap. 4). A list of candidate giant planets in the habitable zone that could host exomoons was recently compiled by Vera Dobos of the University of Groningen and her collaborators [54]. Excluded from this list were brown dwarfs. The mass of a hypothetical exomoon was assumed to be between one-hundredth and one-tenth of the planetary mass, with limits placed on the moon's mass and orbital distance from its planet to ensure stability of its orbit. Since the assumed mass of this moon is larger than the mass of moons that might form by accretion, like the Galilean moons of Jupiter, this only includes large moons formed by collision or by capture, processes whose true probabilities are highly uncertain. This paper is not really asking how likely is it that a particular star hosts an exomoon. Rather, it poses the question: if it is possible that a large exomoon exists around a particular star, how likely is it that it could be habitable?

This is assessed by running many simulations of a simple model of exomoon habitability, using different combinations of plausible physical parameters. In the case of Harriot, most simulations were too hot, but 35% were habitable. Most of the planets on this list are distant and not within our 10 parsec limit, however. The value of this list is that it narrows down the search for habitable exomoons to planets where they could actually exist. Also, M stars are not completely excluded as hosts for habitable exomoons, as Lalande 21185 c and Proxima Centauri c are on this list, although as both planets are cold, they have a much lower chance of hosting a suitable moon. Having said this, smaller moons, up to Titan or Mercury size, might be found in stable orbits around some giant planets circling M stars [55]. Under some special circumstances, these smaller moons could be habitable [52].

Habitable exomoons are not the only unusual type of habitable world that could exist in this sector. Slightly further out, at 38 parsecs, is one of the closer Kepler planets, K2-18 b [56]. The "K2" designation relates to an extension of the original Kepler mission. In addition to its primary mission that focused on the Kepler field of view (Chap. 3), after an equipment malfunction a secondary mission was completed that observed other parts of the sky [57]. K2-18 b circles an M star and its mass and radius are both reasonably well known. It is a fascinating warm sub-Neptune, because it could be the archetype of a new class of planets, the Hycean worlds [58].

An Hycean world has a temperate ocean that lies beneath an atmosphere dominated by hydrogen, instead of nitrogen like Earth. Recent spectra taken with the JWST clearly show that K2-18 b has an atmosphere containing methane, carbon dioxide, and water vapor, as constituents of an atmosphere that is also likely rich in hydrogen [59]. Ammonia was not detected, and a possible explanation is that because ammonia dissolves in water, it has been removed from the atmosphere by a large water ocean. There was also a tentative detection of dimethyl sulphide, a compound usually produced on Earth by biological processes in the ocean. If this signal is confirmed, it could have profound implications for the potential for life on this planet.

Climate simulations of K2-18 b are under way, using a number of general circulation models. The project is named (take a deep breath) Comparing Atmospheric Models of Extrasolar Mini-Neptunes Building and Envisioning Retrievals and Transits (CAMEMBERT) [60]. This cheesy acronym promises to provide new and improved estimates of the climate of K2-18 b. For now, there are estimates of its surface climate from one-dimensional climate models, models that do not incorporate winds or currents. These estimates tend to suggest that the "ocean" of K2-18 b is hot, under high pressure, and may not even be liquid [61]. Instead, it might be a supercritical fluid (see Chap. 3). The main problem affecting habitability on Hycean worlds is their vulnerability to runaway greenhouse, which is exacerbated for a world with lots of water vapor. This might push the inner boundary of the habitable zone for an Hycean world out to 1.6 a.u. in our solar system, to where a planet would only receive about 40% of the amount of solar radiation that Earth receives [62]. Since K2-18 b receives about 90% of Earth's radiation, this is well above the estimated runaway greenhouse limit for an Hycean world. Still, there is that detection of dimethyl sulphide. We shall see.

It would be careless not to mention the Gemini twins, Castor and Pollux, who live in this sector. At just over 10 parsecs away, Pollux is the nearest giant, a K star less than a billion years old that shines with about 30 times the brightness of our Sun. It has a planet, Thestias, a hot Jovian or super-Jovian, about which little is known. Castor is not that much further away, at about 16 parsecs, and actually consists of six young stars, three pairs of close binaries comprising two main sequence A stars and four red dwarfs. Stellar systems like Castor with six stars are rare, with only a few other examples known. In Greek mythology, while Castor and Pollux were twins and had the same mother, Leda, they had different fathers. This is biologically possible and is known as heteropaternal superfecundation [63]. While Leda's legitimate husband was a king of Sparta, one night Zeus took the form of a swan, snuck into her bed, and impregnated her as well. The myth of Leda and the Swan has been used as a subject by many artists and writers, from Roman times through Michelangelo to the TV series *Orphan Black* (2013–2017) [64, 65], where "Project Leda" is the name of a sinister cloning experiment. The twin stars are also seen as brothers in some Australian First Nations traditions [66].

Finally, at 24 parsecs is the system containing the nearest B star, Regulus [67], along with its three smaller companion stars. Its name means "little king" and it was given to it by Polish astronomer Nicolas Copernicus, the originator of the idea that the Earth orbits the Sun instead of the other way around. Regulus is in the zodiacal constellation Leo, the lion, and the Greeks identified Leo with the dreaded Nemean lion killed by Hercules as one of his twelve labors [68]. It is one of the oldest constellations and the ancient Egyptians knew how important it was [69]. The Sun was in Leo at the time of the flooding of the Nile, a vital resource for their agriculture [70].

Procyon Sector Summary

Superflares, Luyten and his star, the lava world Janssen, the possibility of habitable exomoons, and Hycean worlds; this sector has the usual rich complement of places to visit. Like the Summer Triangle formed by Altair, Vega and Deneb (see Chap. 3), there is also a Winter Triangle, formed by Sirius, Procyon and Betelgeuse. Sirius and Procyon we have already encountered and are both nearby main sequence stars. Betelgeuse is not nearby and is nowhere near the main sequence. In fact, it is so far off the main sequence that… well, we can talk about that in the next chapter.

References

1. Kelley DH, Milone EF, Aveni AF (2011) Exploring ancient skies: a survey of ancient and cultural astronomy. Springer, New York
2. Liebert J et al (2013) The age and stellar parameters of the Procyon binary system. Astrophys J 769(1):10. https://doi.org/10.1088/0004-637X/769/1/7
3. Rogers JH (1998) Origins of the ancient constellations: I. The Mesopotamian traditions. J Br Astronom Assoc 108:9–28
4. MacDonald J (1998) The Arctic sky: Inuit astronomy, star lore, and legend. Royal Ontario Museum/Nunavut Research Institute, Toronto, p 72
5. Wikipedia (2023) Known space. Accessed 2 Nov 2023
6. Upgren AR (1995) Willem Jacob Luyten (1899-1994). Publ Astronom Soc Pac 107(713):603
7. Pozuelos FJ, Suárez JC, de Elía GC et al (2020) GJ 273: on the formation, dynamical evolution, and habitability of a planetary system hosted by an M dwarf at 3.75 parsec. Astronom Astrophys 641:A23
8. Chambers JE (1999) A hybrid symplectic integrator that permits close encounters between massive bodies. Mon Not Roy Astronom Soc 304(4):793–799
9. Ferreira D, Marshall J, O'Gorman PA, Seager S (2014) Climate at high-obliquity. Icarus 243:236–248
10. Kilic C, Raible CC, Stocker TF (2017) Multiple climate states of habitable exoplanets: The role of obliquity and irradiance. Astrophys J 844(2):147
11. Colose CM, Del Genio AD, Way MJ (2019) Enhanced habitability on high obliquity bodies near the outer edge of the habitable zone of Sun-like stars. Astrophys J 884(2):138
12. Wilhelm C, Barnes R, Deitrick R, Mellman R (2022) The ice coverage of Earth-like planets orbiting FGK stars. Planet Sci J 3(1):13
13. Li G, Batygin K (2014) On the spin-axis dynamics of a moonless Earth. Astrophys J 790(1):69
14. Yang J, Boué G, Fabrycky DC, Abbot DS (2014) Strong dependence of the inner edge of the habitable zone on planetary rotation rate. Astrophys J Lett 787(1):L2
15. Snellen IA, Brandl BR, De Kok RJ et al (2014) Fast spin of the young extrasolar planet β Pictoris b. Nature 509(7498):63–65
16. Li J, Jiang JH, Yang H et al (2021) Rotation period detection for Earth-like exoplanets. Astronom J 163(1):27
17. Niven L (1968) The handicapped. In: Neutron star. Ballantine, New York, pp 209–236
18. Bethesda (2023) Starfield. Accessed 4 Nov 2023
19. Starfield Wiki (2023) Luyten's rock. Accessed 4 Nov 2023
20. May A, Dobrijevic D (2022) The Carrington Event: History's greatest solar storm. space.com. https://www.space.com/the-carrington-event. Accessed 5 Nov 2023

21. Lasar M (2012) 1859's "Great Auroral Storm"—the week the Sun touched the Earth. arstechnica.com. https://arstechnica.com/science/2012/05/1859s-great-auroral-stormthe-week-the-sun-touched-the-earth/. Accessed 5 Nov 2023

22. Wallace D (2022) Solar storm knocks out farmers' high-tech tractors—an electrical engineer explains how a larger storm could take down the power grid and the internet. The Conversation. https://theconversation.com/a-large-solar-storm-could-knock-out-the-power-grid-and-the-internet-an-electrical-engineer-explains-how-177982. Accessed 5 Nov 2023

23. Asimov I (1992) Worlds in order. In: The secret of the universe. Doubleday, New York, p 63

24. Schaefer BE, King JR, Deliyannis CP (2000) Superflares on ordinary solar-type stars. Astrophys J 529(2):1026

25. Okamoto S, Notsu Y, Maehara H et al (2021) Statistical properties of superflares on solar-type stars: results using all of the Kepler primary mission data. Astrophys J 906(2):72

26. Loughran TF, Walsh KJE (2014) Climate model simulations of the meteorological effects of superflares. Bull Aust Meteorol Oceanogr Soc 27:27–33

27. Lingam M, Loeb A (2017) Risks for life on habitable planets from superflares of their host stars. Astrophys J 848(1):41

28. ESA (2024) Gaia overview. https://www.esa.int/Science_Exploration/Space_Science/Gaia_overview. Accessed 17 Jan 2024

29. ESA (2019) A history of astrometry—Part III. https://sci.esa.int/web/gaia/-/53198-astrometry-in-space. Accessed 18 Jan 2024

30. Kervella P, Arenou F, Thévenin F (2022) Stellar and substellar companions from Gaia EDR3-Proper-motion anomaly and resolved common proper-motion pairs. Astronom Astrophys 657:A7

31. Clement H (1971) Star light. Ballantine, New York

32. Benford G (1984) Across the sea of suns. Simon and Schuster/Timescape, New York

33. Gatewood G (1996) Lalande 21185. Bull Am Astronom Soc 28:885

34. Beatty TG (2022) The detectability of nightside city lights on exoplanets. Mon Not Roy Astronom Soc 513(2):2652–2662

35. Star trek: the next generation (1987) imdb.com. https://www.imdb.com/title/tt0092455/?ref_=nv_sr_srsg_0_tt_8_nm_0_q_star%2520trek%2520next%2520. Accessed 12 Jan 2024

36. Bowens-Rubin R, Murphy JMA, Hinz PM et al (2023) A Wolf 359 in sheep's clothing: Hunting for substellar companions in the fifth-closest system using combined high-contrast imaging and radial velocity analysis. Astronom J 166(6):260

37. Bonfils X (2017) A temperate exo-Earth around a quiet M dwarf at 3.4 parsecs. Astronom Astrophys 613:A25. https://doi.org/10.1051/0004-6361/201731973

38. Miguel Y, Cridland A, Ormel CW et al (2020) Diverse outcomes of planet formation and composition around low-mass stars and brown dwarfs. Mon Not Roy Astronom Soc 491:1998–2009. https://doi.org/10.1093/mnras/stz300

39. Herath M, Gunesekera S, Jayaratne C (2021) Characterizing the possible interior structures of the nearby exoplanets Proxima Centauri b and Ross-128 b. Mon Not Roy Astronom Soc 500:333–354. https://doi.org/10.1093/mnras/staa3110

40. Kite ES, Ford EB (2018) Habitability of exoplanet waterworlds. Astrophys J 864(1):75

41. McIntyre SRN (2022) Tidally driven tectonic activity as a parameter in exoplanet habitability. Astronom Astrophys 662:A15

42. Ibid.

43. Quick LC, Roberge A, Mendoza GT et al (2023) Prospects for cryovolcanic activity on cold ocean planets. Astrophys J 956(1):29

44. Barnes R, Mullins K, Goldblatt C et al (2013) Tidal Venuses: triggering a climate catastrophe via tidal heating. Astrobiology 13(3):225–250

45. Bowen W (2021) A quantum hack for microscopes can reveal the undiscovered details of life. The Conversation. https://theconversation.com/a-quantum-hack-for-microscopes-can-reveal-the-undiscovered-details-of-life-161182. Accessed 16 Jan 2024

46. Rasmussen KC, Currie MH, Hagee C et al (2023) A nondetection of iron in the first high-resolution emission study of the lava planet 55 Cnc e. Astronom J 166(4):155

47. Hu R, Bello-Arufe A, Zhang M (2024) A secondary atmosphere on the rocky exoplanet 55 Cancri e. Nature. https://doi.org/10.1038/s41586-024-07432-x

48. Madhusudhan N, Lee KK, Mousis O (2012) A possible carbon-rich interior in super-Earth 55 Cancri e. Astrophys J Lett 759(2):L40

49. Musser G (2010) Earth-like planets may be made of carbon. Sci Am 302(1) https://www.scientificamerican.com/article/a-large-lump-of-coal/

50. Hill ML, Bott K, Dalba PA et al (2023) A catalog of habitable zone exoplanets. Astronom J 165(2):34

51. Bloom TF (2016) Borrowed perceptions: Harriot's maps of the Moon. J Hist Astronom 9(2):117–122. https://doi.org/10.1177/002182867800900203

52. Arnscheidt CW, Wordsworth RD, Ding F (2019) Atmospheric evolution on low-gravity waterworlds. Astrophys J 881(1):60

53. Dobos V, Haris A, Kamp IEE, van der Tak FFS (2022) A target list for searching for habitable exomoons. Mon Not Roy Astronom Soc 513(4):5290–5298. https://doi.org/10.1093/mnras/stac1180

54. Ibid.

55. Martinez-Rodriguez H, Caballero JA, Cifuentes C et al (2019) Exomoons in the habitable zones of M dwarfs. Astrophys J 887(2):261

56. Sarkis P, Henning T, Kürster M et al (2018) The CARMENES search for exoplanets around M dwarfs: A low-mass planet in the temperate zone of the nearby K2-18. Astronom J 155(6):257

57. Howell SB, Sobeck C, Haas M et al (2014) The K2 mission: characterization and early results. Publ Astronom Soc Pac 126(938):398

58. Madhusudhan N, Piette AA, Constantinou S (2021) Habitability and biosignatures of Hycean worlds. Astrophys J 918(1):1
59. Madhusudhan N, Sarkar S, Constantinou S et al (2023) Carbon-bearing molecules in a possible Hycean atmosphere. Astrophys J Lett 956(1):L13
60. Christie DA, Lee EK, Innes H et al (2022) CAMEMBERT: A mini-Neptunes general circulation model intercomparison, protocol version 1.0. A CUISINES Model Intercomparison Project. Planet Sci J 3(11):261
61. Scheucher M, Wunderlich F, Grenfell JL et al (2020) Consistently simulating a wide range of atmospheric scenarios for K2-18 b with a flexible radiative transfer module. Astrophys J 898(1):44
62. Innes H, Tsai SM, Pierrehumbert RT (2023) The runaway greenhouse effect on Hycean worlds. Astrophys J 953(2):168
63. Wikipedia (2024) Superfecundation. Accessed 26 Jan 2024
64. Wikipedia (2024) Leda and the Swan. Accessed 26 Jan 2024
65. Wikipedia (2024) Orphan Black. Accessed 26 Jan 2024
66. Hamacher D (2017) Kindred skies: ancient Greeks and Aboriginal Australians saw constellations in common. The Conversation. https://theconversation.com/kindred-skies-ancient-greeks-and-aboriginal-australians-saw-constellations-in-common-74850/. Accessed 20 Mar 2024
67. Wikipedia (2024) Regulus. Accessed 21 Mar 2024
68. Pasachoff JM (2006) Stars and planets. Houghton Mifflin, Boston
69. Wikipedia (2024) Leo (constellation). Accessed 23 Mar 2024
70. Lea R (2022) Leo constellation: facts, location, and stars of the lion. space.com. https://www.space.com/16845-leo-constellation.html. Accessed 23 Mar 2024

7

The Bull, the Warrior and the Queen
The Eta Cassiopeiae Sector: 0–6 h, 0 to +90°

Betelgeuse, the second-brightest star in Orion, is a red supergiant about 170 parsecs away, so nowhere near the known galaxy. This is a very good thing, as Betelgeuse could explode at any moment.

Now when we say "at any moment", what astronomers really mean is "sometime in the next few hundred thousand years", which is just about tomorrow on the time scales that they are used to dealing with [1]. As a result, it would be an exaggeration to say that astronomers are nervous about Betelgeuse. The star is so far away that even if it exploded, the only real effects would be that it would suddenly become by far the brightest star in the night sky and would be easily visible during the day as well. There would also be a temporary increase in UV lasting about an hour [2]. So astronomers are curious about Betelgeuse, not nervous (Fig. 7.1).

This curiosity greatly increased a few years ago when Betelgeuse suddenly became much dimmer. On December 8, 2019, a message was posted on The Astronomer's Telegram, an internet-based service that publishes alerts and notifications about new astronomical observations [3]. Two astronomers from Villanova University, Edward Guinan and Richard Wasatonic, along with experienced amateur observer and telescope builder Thomas Calderwood, reported that in the previous 2 months, Betelgeuse had undergone an almost unprecedented decrease in brightness. Known as the Great Dimming, it immediately raised fears in the news media that the star was about to explode. Later, a more prosaic explanation was provided: Betelgeuse had ejected some hot gas that later condensed into a dust cloud, temporarily obscuring the star [4].

Nevertheless, the possibility that Betelgeuse could undergo a supernova explosion in our lifetimes is not zero. We would even get a few hours advance

K. J. E. Walsh, *Planets of the Known Galaxy*, Science and Fiction,
https://doi.org/10.1007/978-3-031-68218-6_7

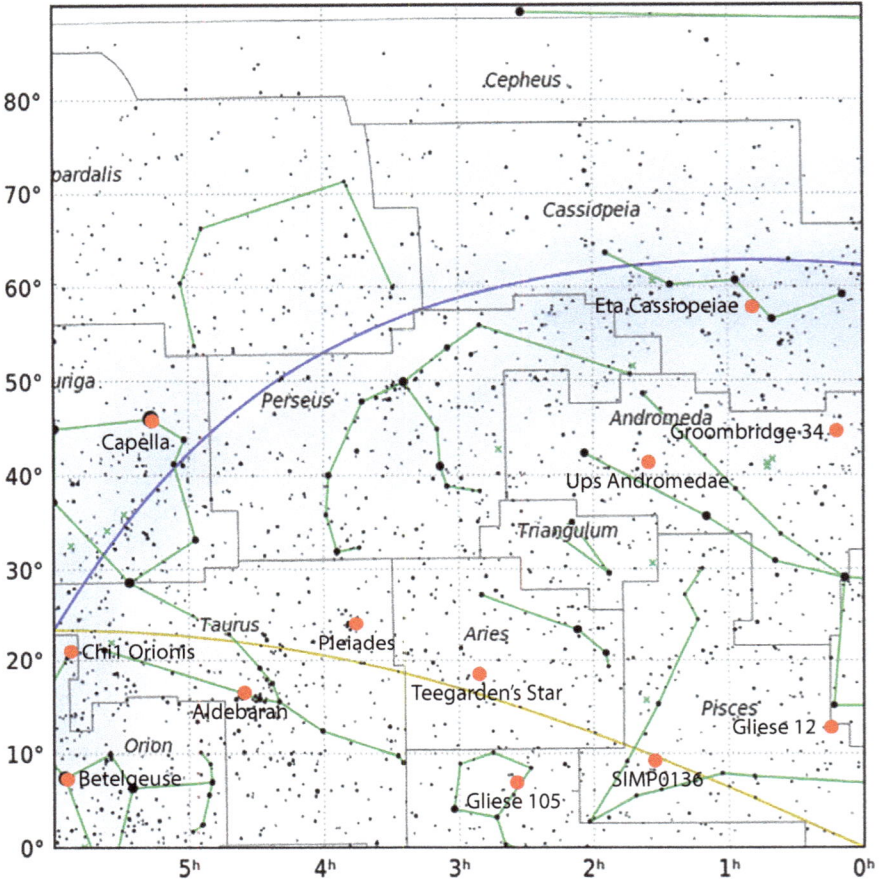

Fig. 7.1 Map of the Eta Cassiopeiae sector. Blue shading indicates the "Milky Way", the most star-packed part of the galaxy, while the blue (darker) line is the galactic equator, an imaginary circle running through the Milky Way that slices the galaxy in half. The orange (lighter) line is the path of the Sun's location through the seasons (indicating the "zodiac" constellations). Larger circles indicate brighter stars, while locations mentioned in this chapter are indicated by red circles. Image produced using StarCharter (https://github.com/dcf21/star-charter, accessed on the 2nd of July 2024. © 2020, Dominic Ford) software under the GNU general public license

warning. One of the first signs that the star has gone supernova would be an enormous outward flux of neutrinos, almost massless sub-atomic particles that are formed in great numbers in such explosions. These can be detected by the Super-Kamiokande neutrino observatory, consisting of a large number of detectors immersed in very clear water in a cavern about 1000 m underground in a mine in Japan [5]. Once the neutrino burst was observed, a few hours later the star would greatly increase in brightness, and the supernova event would have begun.

Betelgeuse is in the constellation Orion, one of the easiest to recognize in the night sky because it actually does look a little like the outline of a human figure, wearing a belt with what might be a sword attached to it (Fig. 7.2).

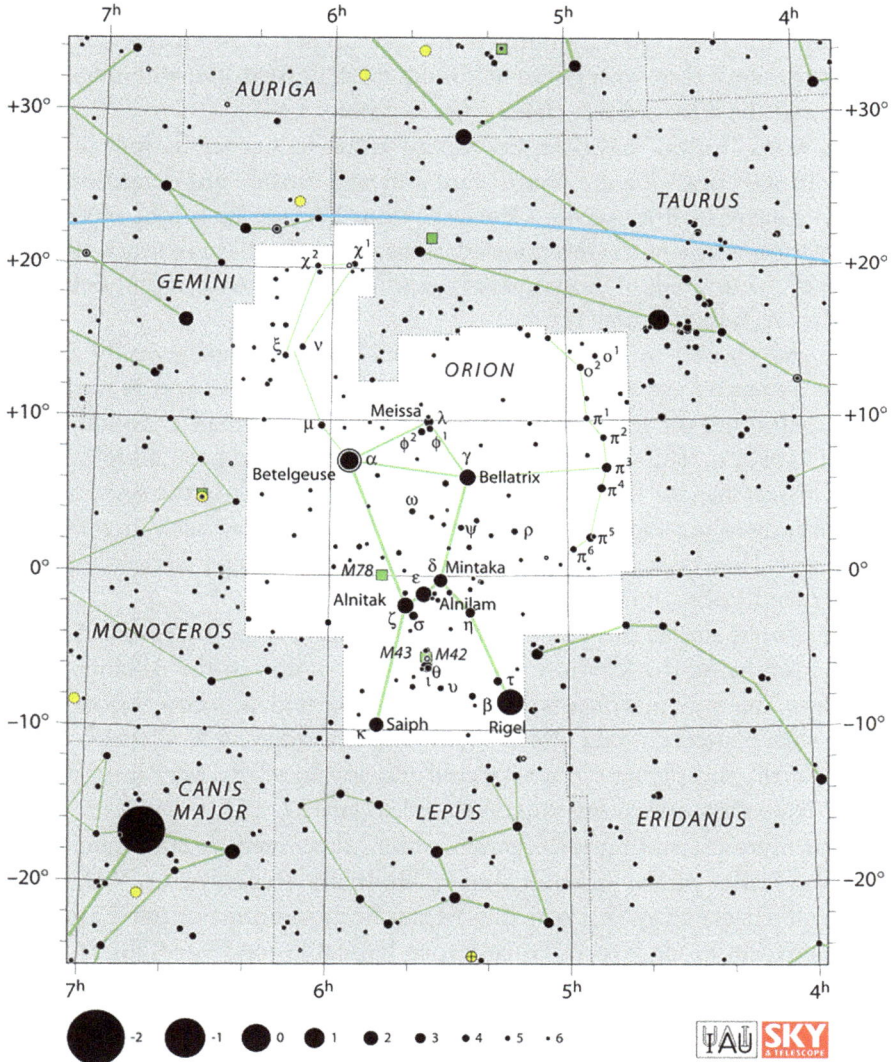

Fig. 7.2   Sky map of Orion. Size of circles indicates stellar magnitude (brightness), with negative values being brighter. Reprinted under CC-BY-4.0 license from https://www.iau.org/public/themes/constellations/ . © 2024, International Astronomical Union

This is the way that the constellation is represented throughout ancient Europe and the Middle East [6]. Other cultures see Orion differently. The Chinese split Orion into a few sub-constellations, including the belt of Orion as the "three stars" [7]. For the Lakota people over the North American plains, the bottom half of Orion was the arm of a Lakota chieftain, put there by the gods as punishment for selfishness [8]. For the Kokatha people of Australia, Orion was the egotistical hunter Nyeeruna, whose periodic generation of fire magic explained the observed variability of the brightness of Betelgeuse. In noting these variations, they were better astronomers than Aristotle, who once baldly stated that the stars did not change and did not vary [9].

The word "Orion" has been used many times in science fiction, but since the main stars of Orion are very distant and well outside our definition of the known galaxy, we'll move on. One last point: because its bright stars are far away, Orion is one of the few constellations that will still look much the same tens or even hundreds of thousands of years from now, provided that Betelgeuse is still there, which it may not be.

The star chosen to represent this sector is Eta Cassiopeiae, only about 6 parsecs away and consisting of a pair of main sequence G and K stars. They circle each other on an eccentric orbit at a distance as small as 36 a.u. and as large as 106 a.u. This 5-billion-year-old system has long been a high-priority target of searches for habitable zone planets and was one of the systems listed as such in Stephen Dole's *Habitable Planets for Man*, way back in 1964 [10]. No planets have been found yet. The two stars are well separated, so terrestrial planets could exist in both their habitable zones [11].

The nearest star in this sector with a confirmed planet is Groombridge 34. Apart from being the location of a small space station in the Alliance-Union Universe, the system consists of two well-separated M stars about 90 a.u. apart, with planets circling one of them, Groombridge 34 A. Planet b is a hot super-Earth, and planet c is a very cold sub-Jovian with a very long orbit, at about 7600 days one of the longest found so far [12, 13].

Even more interesting is Teegarden's Star, discovered in 2003 and named after the leader of the discovery team, Bonnard Teegarden of NASA [14]. Only 3.8 parsecs away, this very dim M star hosts not one but two Earth-sized planets within or close to its conservative habitable zone (see Table 7.1) [15, 16]. Teegarden's Star b has the highest known Earth Similarity Index [17]. This planet also gets a high score from SEPHI, with a good but not perfect chance of retaining liquid water on its surface, due to the planet being close to the inner edge of the star's habitable zone. Planet c is a lot cooler and has a higher chance of retaining liquid water, but where both worlds fall down is in the strength of their magnetic fields, which may be weak. Most likely, both

Table 7.1 Planets of Teegarden's Star

Planet	Mass (Earth = 1)	Orbital period (days)	Distance from star (Earth distance from Sun = 1)	Equilibrium temperature (degrees K) (Earth = 255 K)	Type
Teegarden's Star b	>1.16	4.91	0.0259	259–283	Warm or hot Earth-size or super-Earth
Teegarden's Star c	>1.05	11.4	0.0455	196–214	Warm or hot Earth-size super-Earth
Teegarden's Star d	>0.82	26.1	0.0791	148–162	Cold Earth-size

planets b and c are tidally locked. If the planets have retained an atmosphere after the turbulent youth of Teegarden's Star, an initial habitability assessment by Amri Wandel of the Hebrew University of Jerusalem and collaborators suggests that the most habitable regions are on the night side of planet b, but at the day side sub-stellar point of planet c [18]. Assuming a greenhouse effect similar to Earth's and an atmosphere that is able to redistribute some heat from the day side to the night side (that is, an atmosphere thicker than about one-tenth of Earth's), temperatures on the night side of planet b should be less than 300 K, perhaps reaching freezing at the anti-solar point [19]. It would be hotter on the day side, perhaps about 350 K, or far too hot for humans.

This assumes that planet b has avoided runaway greenhouse, which is not at all clear. Colby Ostberg of U.C. Riverside and collaborators point out that planet b actually resides within the Venus Zone [20]. Or does it? Currently, the location of the runaway greenhouse boundary is estimated from theory, and there are no observations of actual extrasolar planetary conditions to back it up. Clouds are not really incorporated into the basic theory, and their effects could modify the results considerably [21]. Climate modeling studies that do include the effect of clouds suggest that the actual runaway greenhouse limit might be closer to stars than the theoretical limit [22]. Since planet b is only just a little closer to its star than the accepted theoretical runaway greenhouse boundary, it would be an excellent test bed for validation of the runaway greenhouse concept. If measurements of its temperature and atmospheric constituents can be obtained, and if they indicate habitable conditions, then the runaway greenhouse boundary might be closer to stars than we think. The planets of Teegarden's star are top priority observing candidates for the JWST, so we should know more soon. There will be challenges, though. If planet b is like Venus and has a thick atmosphere, this would make it difficult to detect the light of its star passing through its atmosphere and thereby analyze its spectrum. A thinner atmosphere would allow light to pass through more easily and so would be more likely to show a spectrum. Once the approximate size and constituents of the atmosphere are determined, models could then be used to estimate the surface temperature and estimate its habitability.

On planet c, Wandel's paper suggests that if it had an Earth-like greenhouse effect, temperatures at the sub-stellar point would be only slightly above freezing. If it had a greenhouse effect twice that of Earth—which of course is entirely possible—the sub-stellar temperature would be closer to 320 K (about 50 °C or 120 °F). This is a big temperature jump and once again shows how important the size of the greenhouse effect is to planetary habitability. In compensation, for this "double-greenhouse" world, more day side regions of the planet would have milder temperatures than for the Earth-greenhouse world.

A more detailed description of Double-Greenhouse World gives some insight into these kinds of planets, even if planet c is not exactly like that. These are worlds that require a greenhouse effect larger than Earth's in order to have temperate conditions somewhere on their surface, because they do not receive enough stellar radiation. If the sub-stellar point on these planets is a large ocean, then it will be very cloudy and stormy there. This cloud may in fact be so dense that it will act to considerably reduce the maximum surface temperature from its predicted value, and since the cloud would be continually replenished by thunderstorms, it would rarely dissipate. On the other hand, if this region were land, there could be a substantial area of very hot desert. Climate simulations suggest that the region of maximum precipitation would be displaced slightly to the east of the sub-stellar point (see Fig. 2.2), depending on rotation rate and the relative positioning of bodies of land and water [23]. Away from this torrid region, conditions would become more temperate, with the most clement parts of the planet on the day side about 45° away from the sub-stellar point. At the terminator, though, mean temperatures would have already fallen below freezing, and ocean regions there would have sea ice. This temperature decrease would continue into the night side, with temperatures at the anti-stellar point about −30 °C (−4 °F). The night side would be full of glaciers and frozen oceans.

As mentioned earlier, simulations indicate that a high water content and a global ocean, frozen or unfrozen, are the most likely outcomes for planets of Earth mass or larger circling M stars [24]. One of the earliest simulations of planetary formation, if not the earliest, was published in 1970 by Stephen Dole [25]. Using a random number generator and some simple rules for how planets might grow and merge, he was able to generate some realistic-looking solar systems—or at least realistic as the science was understood at the time he performed his study. In his simulated solar systems, all of the gas giants ended up in the outer solar system and most of the terrestrial planets were in the inner solar system. Planetary migration was not incorporated into his modeling system because it was unknown at the time, and this can lead to very different results. One of the best-cited papers on simulations of planetary formation is Sean Raymond's 2004 work that not only generates planets but also their varying water contents [26]. They use the MERCURY model (see Chap. 6), whose main advantage is that the actual physics of the gravitational interaction of planetary embryos is incorporated in simplified form, rather than just applying statistical probabilities of how they might interact. The artificial solar systems produced are surprisingly realistic (Fig. 7.3), with planets in the habitable zone ranging from super-Earth water worlds to terrestrial-sized desert planets [27]. Since then, further refinements have been made to this method, to the point where models like this one are being used to try to resolve fundamental questions regarding the way planets form.

Even if they are very watery, planets receiving even less stellar radiation than Double-Greenhouse World could also have habitable temperatures if they possessed a large enough greenhouse effect. Even so, with increasing

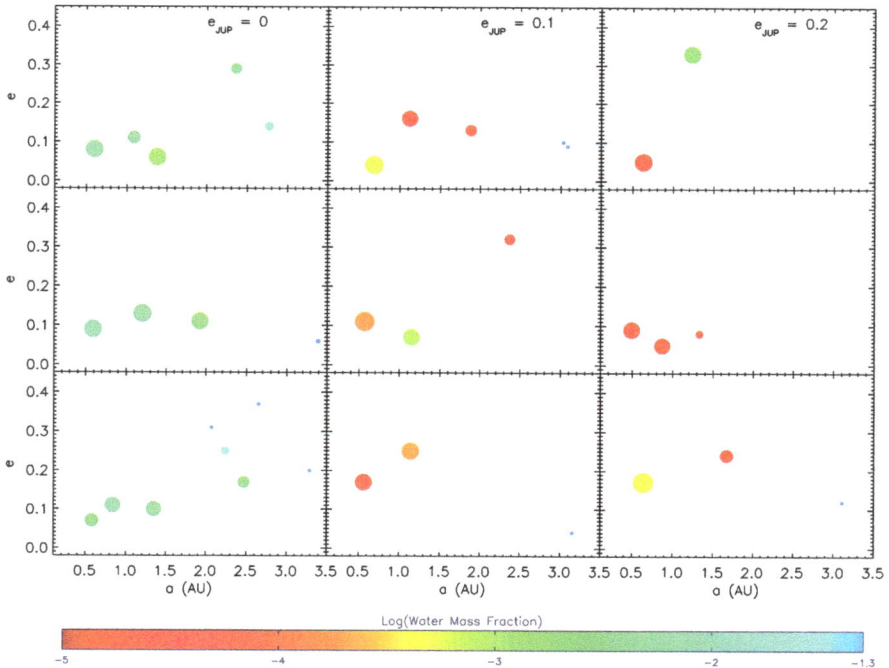

Fig. 7.3 Simulated formation of planetary systems with the same initial conditions except that the eccentricity of Jupiter is varied, from e = 0 in the left column to e = 0.2 in the right column. Colors indicate water content, with high water content blue and low water content red. For comparison, Earth's water mass fraction is roughly yellow on this plot. Reprinted with permission from [26]. © 2004, Elsevier B.V. All rights reserved

distance from the star, at some point the required greenhouse warming would need an atmosphere so thick as to make the planet inhospitable, more like a cold version of Venus than like Earth. Also, the maximum greenhouse limit would be reached (see Chap. 1), making further warming difficult to achieve [28]. Still, it would be good if we had more information on planetary temperatures and atmospheric composition, so that we could get a better handle on these limits.

Teegarden's Star c *How similar to Earth: Likely colder, weaker magnetic field.*
Plausible planet: Double-greenhouse ocean world. Hot, stormy, cloudy weather near the sub-stellar point on the day side, glaciers and frozen oceans on the night side.
Best case scenario: Some temperate land regions on day side.
Worst case scenario: Frozen solid, with a thin or absent atmosphere.
Cultural connection: Double greenhouse world.

*   *   *

Rogue planets and brown dwarfs keep on being discovered as detection techniques improve. The T dwarf SIMP J013656.5+093347 lives in this sector in the zodiacal constellation Pisces, and to avoid wearing out the reader's patience, it will be given the name Simpy. Simpy is only about 13 Jupiter masses, so it is right on the borderline between brown dwarf and super-Jovian [29]. As a young T dwarf, it has a temperature of about 1000 K, and like Luhman 16A, it probably looks purple or magenta. Simpy has a couple of claims to fame. It is nearby, only about 6 parsecs away, and is part of a group of objects called the Carina-Near Moving Group, a very loose collection of stars that have similar ages and directions of movement [30]. It also has auroras, big ones. In 2018, Melodie Kao of CalTech and her collaborators used a very large radio telescope to discover that Simpy had a magnetic field about 1000 times stronger than Jupiter's and about 20,000 times stronger than Earth's [31]. The strength of its magnetic field was deduced from the strength of the high frequency radio signals emitted by its auroras.

As a T dwarf, Simpy also has an extensive atmosphere. Recent work by Johanna Vos of the American Museum of Natural History and her team has used observations of the spectrum of Simpy to conclude that it has a deep, thick layer of iron clouds overlain by higher, patchy clouds of the crystalline mineral forsterite [32]. Forsterite (Mg_2SiO_4) is perhaps not the first potential cloud constituent that springs to mind in this context, but there it is. The infrared radiation output of SIMPY is variable, and this appears to be well explained by the movements of the forsterite clouds across the planet. SIMPY rotates very quickly, once every 2.4 h, so circular storms with clouds should be frequent [33]. Massive convective storms (e.g. thunderstorms) should also be common.

* * *

One of the hazards of the current state of the art of planetary discovery is that it is sometimes difficult to pull sensible answers out of the data. Orbital solutions for a number of planets can be obtained from radial velocity data, but then the solutions simply do not add up when further analysis is applied. For instance, take the planets of the 82 Eridani system, a G star about 6 parsecs away in the Epsilon Eridani sector (Chap. 8). Three planets are confirmed (see Table 7.2), but the currently accepted values of the orbital eccentricities of planets d and e mean that their orbits would cross [34], which is a big planetary no-no. In that case, both orbits would certainly be unstable, and since the 82 Eridani system is at least 5 billion years old and its planets are still there, the true values of the eccentricities of these two planets must be much lower. Part of the problem is the extreme inaccuracy of the current estimates of eccentricity. How then to constrain the orbits to more realistic shapes? In

Table 7.2 Planets of 82 Eridani

Planet	Mass (Earth = 1)	Orbital period (days)	Distance from star (Earth distance from Sun = 1)	Equilibrium temperature (degrees K, assuming albedo = 0.3) (Earth = 255 K)	Type
82 Eridani g (unconfirmed)	>1.0	11.9	0.095	766	Hot Earth-size
82 Eridani b	>2.8	18.3	0.127	663	Hot super-Earth
82 Eridani c (unconfirmed)	>2.5	43.2	0.225	498	Hot super-Earth
82 Eridani d	>3.5	88.9	0.364	391	Hot super-Earth
82 Eridani e	>4.8	147.0	0.509	331	Hot super-Earth
82 Eridani f (unconfirmed)	>10.3	331.4	0.875	252	Warm sub-Neptune
82 Eridani h (predicted)	1–8	611	0.99[a] 1.31	206	Warm super-Earth

[a]This is an update to Table 3 of Basant et al. [35]. Based on the assumed stellar mass and the orbital period of planet h, this distance should be about 1.31 a.u.

2022, Ritvik Basant, Jeremy Dietrich and Daniel Apai of the University of Arizona performed a wide-ranging study of the possible characteristics of this planetary system. They used the DYNAMITE model (DYNAmical Multi-planet Injection TEster—another acronym that I am sure you will agree is not at all contrived) [35]. In this model, distributions of planetary characteristics from the Kepler data were matched to the 82 Eridani system and the probabilities of various configurations were estimated, and then used to predict the most likely location of any additional planets. Then, the stability of the resulting planetary system was checked by simulations of the orbits of the planets and their gravitational interactions. They concluded that there was good evidence for the existence of all six of the planets listed in Table 7.2, plus one more. Both this predicted planet, planet h, and planet f would be in the conservative habitable zone, while the other five planets are too hot. For this system to be gravitationally stable, though, much lower eccentricities have to be assumed than those published. The authors point out that better observations are needed (as they always are…).

Fictional planets around 82 Eridani include Rustum, found in Poul Anderson's *Orbit Unlimited* (1961) and in the later short story collection *New America* (1982) [36, 37]. This beautiful, large but hostile planet is hotter than Earth and has a bigger ocean. A persistent cloud layer separates the unpleasantly thick atmosphere of the lowland regions from the more habitable, mostly desert uplands about eight kilometers above sea level. It has 62-h days and is circled by two moons, the larger, copper-colored one appearing in the sky of Rustum to be about twice the diameter of Earth's Moon in our sky. The planet's gravity, about 25% greater than Earth's, makes accidental falls very hazardous. As usual in Anderson's work, the description of the planetary environment is very vivid and detailed, especially its alien ecosystem. There is a tree that does not have solid leaves but instead a lacework pattern. The giant spearfowl soars on the dense lowland air and is big enough to kill a man. The plume oak, the gimtree, the bower phoenix, the singing lizard: these all live in the Rustum forest, where the heavy wind rustles the leaves and creates a sound unlike anything heard in autumn in North America. On this planet, the human colonists are trying to establish a free society on Rustum but run headlong into both the alien environment and conflicting political visions of the colony's future. More recently, Stephen Baxter's *Ark* (2009) takes us to Earth II, a world spinning on its side, giving it dramatic seasons that make most of the planet unviable. Climate model simulations suggest that such planets are unlikely to have polar ice caps, and neither does Earth II. Planets can be tilted into almost any orientation by giant impacts during their formation, so this scenario is plausible, although as discussed in Chap. 6, sideways-spinning worlds may be more habitable than previously thought.

The discovery of many compact stellar systems crammed full of planets, like that of 82 Eridani, was one of the surprises in the exploration of the known galaxy. But in some systems, planets might not even be able to form. A companion star that is close enough to the primary could disrupt the formation of planets. Despite this, a fairly large number of planets have been discovered that orbit stars that are in binary systems [38]. Most orbit one of the binary stars, but some have been found on orbits circling both, like the planet Tatooine in *Star Wars* [39]. This concept is illustrated by the Gliese 105 system. About 7 parsecs away, it consists of three stars. The biggest is a K star, and about 1200 a.u. away is an M star, but there is also another M star on a very eccentric orbit around the K star with an average distance of about 17 a.u. A 2012 assessment of the stability of planetary orbits in this system by Luisa Jaime of the National Autonomous University of Mexico and collaborators suggests that planets around the K star would have to orbit closer than about 1 a.u. to be on stable orbits, although the orbital elements of the system have been updated a little since that study was published [11, 40]. Since the center of the habitable zone of a K3 star is closer than 1 a.u., orbital stability considerations might not rule out a habitable planet in this case. In contrast, a planet would have to be at least 50 a.u. away from these stars to have a stable, Tatooine-like orbit around both of them. A system in this sector where this issue might prevent a habitable planet is the Chi¹ Orionis system about 9 parsecs away, consisting of a G0 and an M star orbiting within a few a.u. of each other. Orbital stability analysis implies that planets of the G0 star would have to be closer than about 1 a.u., and this might be too close to the inner edge of its habitable zone [11]. While the system is still young, less than 500 million years old, it may never host an Earth-like planet [41].

Further Afield

Just outside the 10 parsec limit is the bright star Capella, a four-star system that includes two giant stars separated by only 0.74 a.u. Despite lots of mentions in science fiction over the years, this 600-million-year-old system has no known planets and because its main stars have already gone giant, it is not a high priority for searches for habitable planets. Also just outside the 10 parsec limit is Gliese 12 b, discovered in 2024 [42]. It is just about the nearest Earth-sized planet with a reasonable equilibrium temperature, although at 315 K (42 °C; 107 °F) it is warm and runaway greenhouse has not yet been excluded.

It is an ideal candidate for atmospheric spectroscopy, though, so we will probably know more soon.

In this sector is one of the first extrasolar planets to be discovered, Upsilon Andromeda b, found in 1996 circling its F star about 13 parsecs away [43]. The planet is roughly the mass of Jupiter and is close to its star, so its radial velocity is easily measured. It was discovered by the team of Paul Butler and Geoff Marcy, whose revolutionary new observation techniques were among the driving forces behind the explosion of extrasolar planet detections in the 1990s. Later, for reasons unrelated to astronomy, Marcy received the dubious honor of being the first member of the prestigious U.S. National Academy of Sciences to be expelled [44].

Upsilon Andromeda was the first multiple star system found to host more than one planet. These worlds circle Upsilon Andromeda A (star B is an M star at least 750 a.u. away). This system appears to be full of outsized worlds (Table 7.3). Planet b, named Saffar after an eleventh-century Arab astronomer, has an atmosphere, is very hot, has a day-night temperature difference of about 1000 K and extreme winds to go with it [45]. Planet d is called Majriti, also named after a medieval Arab scientist. It is in the outer part of the habitable zone and is large enough that it could host an exomoon of substantial size [46]. The planetary orbits of this system are bizarre in that their inclinations do not line up with each other. This is quite different from our solar system, whereby the planetary orbits have modest inclinations with respect to each other. The largest inclination is Mercury's, whose orbit is inclined to Earth's orbit by only about 7° (the orbit of the dwarf planet Pluto is inclined at 17°). In contrast, the inclinations of the planets b, c, and d of Upsilon Andromeda A are 25, 7, and 24° respectively [47, 48]. That's a big difference for such large planets.

Table 7.3 Planets of Upsilon Andromeda

Planet	Mass (Earth = 1)	Orbital period (days)	Distance from star (Earth distance from Sun = 1)	Equilibrium temperature (degrees K, assuming albedo = 0.3) (Earth = 255 K)	Type
Ups And b (Saffar)	541	4.6	0.059	1391	Very hot super-Jovian
Ups And c (Samh)	4445	241.3	0.822	373	Hot super-Jovian or brown dwarf
Ups And d (Majriti)	3259	1276.5	2.55	212	Warm super-Jovian

The orbit of Majriti is also eccentric, with a closest approach to its star of about 1.7 a.u. and a greatest distance of about 3.4 a.u. This is enough to give a big seasonal variation in incoming radiation over its stellar year of 3.5 standard years. At closest approach, its equilibrium temperature is about 256 K, or about the same as Earth's, while at furthest distance it is about 184 K, or about seventy degrees C colder than Earth. What would the seasons be like on a hypothetical large exomoon of Majriti? To simplify things, we assume that the exomoon has zero axial tilt relative to its star and therefore seasonal variations in temperature would be caused only by the variations in its distance from its star. In this case, the maximum incoming stellar radiation would always occur at its equator. Nevertheless, when the planet is at its greatest distance, incoming radiation even at the equator would be low, less than a third of maximum terrestrial values, or about the same as in northern Europe during winter. Radiation amounts at other latitudes would be even less, so it is difficult to see how the planet could avoid a worldwide freeze. In contrast, at closest approach the maximum incoming radiation is about the same as the terrestrial amount. Radiation would be similar to terrestrial values for only a few months of its orbit, so that would probably be only enough to guarantee a sharp, tropical-like summer in the equatorial regions, rather than making the planet uninhabitably hot. While this seasonal variation in temperature is extreme, one can speculate that any life on a hypothetical habitable moon of Majriti could adapt to the long winter by doing what some terrestrial species do: hibernate.

The Andromeda constellation where these planets live is the subject of a famous Greek myth. Andromeda was the daughter of the king of Aethiopia, what the ancient Greeks called the part of Africa south of the Sahara Desert. Her mother was Cassiopeia, who boasted of her beauty and that of her daughter. Greek gods have a tendency to dislike braggarts, so in response Poseidon sent a sea monster to attack the coast of Aethiopia. The king consulted an oracle who said that the sea monster could only be satisfied if the king were to sacrifice his daughter to it [49]. Accordingly, the king, being a practical politician, chained his naked daughter to a rock on the coast and waited for the monster to claim his prize. Luckily, who should be passing by but the Greek hero Perseus, who killed the monster with the same diamond sword that he used to kill Medusa [50, 51]. Later, Perseus and Andromeda married and founded the dynasty that ruled the Greek city-state of Mycenae for some centuries [52]. A later version based loosely on this tale was the film *Clash of the Titans*, a sword-and-sandal epic that was one of the surprise hits of 1981 [53]. A remake was also one of the (very) surprise hits of 2010 [54].

Also close by in this sector is Aldebaran, the brightest star in the zodiacal constellation Taurus (the bull), a K giant shining brightly about 20 parsecs away. Its name means "the follower" in Arabic, since it follows the Pleiades in the night sky as it appears to rotate above us. In *The Lensman* series by E. E. "Doc" Smith, some of the action is set in this stellar system. Simply stated, this series is about the adventures of a galactic police force with some members who have special mental abilities. This saga is chock full of non-stop space opera derring-do, but also contains many vivid descriptions of planetary environments. In the third book of the series, *Galactic Patrol*, the planet Aldebaran I, the location of a pirate base, is described as barren, lifeless, and covered in extinct volcanoes and jagged rock [55, 56]:

> Mountain-side and rocky plain, crater-wall and valley floor, alike and innumerably were pockmarked with sub-craters and with immensely yawning shell-holes, as though the whole planet had been throughout geologic ages the target of an incessant cosmic bombardment.

In the same book, we encounter the planet Trenco, whose bizarre and capricious meteorology is a source of great frustration to visitors. Its oceans are made of a liquid that boils easily, causing them to evaporate during the day and condense at night in terrible downpours. Luckily for weather forecasters, the amount of rain in the equatorial regions is the same every night: forty-seven feet and five inches, or about 14,000 mm. That's one serious rainstorm. As well, there is ferocious lightning, and the wind is so strong that much of the planet's surface has been scoured flat. Such a hostile world would be totally avoided if the planet were not the only source of thionite, a highly addictive drug that fuels a vast galactic organized crime network.

Nearby is the star cluster the Pleiades (Fig. 7.4). Visible in the Northern Hemisphere winter sky, this collection of young, hot B stars about 135 parsecs away has many legends told about it. According to the Western Mono of the First Nations people of North America, the stars represented a group of wives who were too fond of eating onions and so were expelled from their homes by their angry husbands [57]. The Kiowa tell the tale of seven girls who were camped at a stream and were suddenly attacked by bears. The girls leaped onto a rock and prayed to it for protection. As the bears scratched the sides of the rock, it grew taller and taller, until the girls ended up in the stars [58]. The rock with scratches on its sides is now known as Devil's Tower, Wyoming. This lofty butte was also the iconic location of the first UFO landing in the epic Steven Spielberg film *Close Encounters of the Third Kind* (1977) [59].

Finally, also visible in this part of the sky is the Andromeda Galaxy. The nearest big galaxy to the Milky Way, it is about 765,000 parsecs away and yet

Fig. 7.4 The Pleiades star cluster. Reprinted from Hubblesite (https://hubblesite.org/contents/media/images/2004/20/1562-Image.html?Topic=104-stars-and-nebulas&keyword=pleiades, accessed on the 8th July 2024). © 2004, NASA, ESA and AURA/Caltech. Public domain

can still be seen by the naked eye, as a kind of cloudy smudge. About the same size as our own galaxy, it contains hundreds of billions of stars. We will become intimately acquainted with it about 4 billion years from now, as it is expected to collide with our own Milky Way [60]. Until then, the Andromeda Galaxy serves as a reminder of the vastness of our own galaxy, of which the known galaxy is only a tiny corner.

Eta Cassiopeiae Sector Summary

A final word about supernovae. While Betelgeuse is too far away to do any major damage to Earth, it is not impossible that other, closer exploding stars might have had bigger effects in the past. There is some evidence in the geologic record that extinctions on Earth at the end of the Devonian period (about 350 million years ago) could have been caused by supernova events [61]. Other work suggests that supernovae occurring closer than about 10–20 parsecs could cause a mass extinction event that would kill so much phytoplankton in the ocean that the marine ecosystem would crash [62, 63]. These might occur about twice every billion years, odds that most human beings would be happy to live with for now. But the number of planets in the galaxy is big, and this increases the odds that somewhere out there, a few perfectly habitable planets have been wrecked by a supernova. Science fiction writers have long recognized this possibility, and perhaps the most moving example is Arthur C. Clarke's short story "The Star" [64]. In this tale, an expedition from Earth to another stellar system, led by a Jesuit astrophysicist, finds the blasted remnants of a civilization. Only later does it become appalling clear to the priest exactly which star exploded: star of wonder, star of night, star with royal beauty bright, the star of the Magi.

References

1. Dolan MM, Mathews GJ, Lam DD et al (2016) Evolutionary tracks for Betelgeuse. Astrophys J 819(1):7
2. Wheeler JC, Chatzopoulos E (2023) Betelgeuse: a review. Astronom Geophys 64(3):3–11
3. Guinan EF, Wasatonic RJ, Calderwood TJ (2019) The fainting of the nearby red supergiant Betelgeuse. The Astronomer's Telegram, 8 Dec 2019. https://astronomerstelegram.org/?read=13341. Accessed 1 Feb 2024
4. Montargès M, Cannon E, Lagadec E et al (2021) A dusty veil shading Betelgeuse during its Great Dimming. Nature 594:365–368. https://doi.org/10.1038/s41586-021-03546-8
5. Machado LN, Abe K, Hayato Y et al (2022) Pre-supernova alert system for Super-Kamiokande. Astrophys J 935(1):40
6. Ridpath I (2018) Star tales. Lutterworth, Cambridge, p 132
7. Wikipedia (2024) Orion in Chinese astronomy. Accessed 3 Feb 2024
8. National Earth Science Teachers Association (2024) Constellation of the hand. https://www.windows2universe.org/?page=/mythology/hand_orion.html. Accessed 3 Feb 2024
9. Hamacher D (2017) Stars that vary in brightness shine in the oral traditions of Aboriginal Australians. The Conversation. https://theconversation.com/stars-that-vary-in-brightness-shine-in-the-oral-traditions-of-aboriginal-australians-85833. Accessed 4 Feb 2024
10. Dole S (2007) Habitable planets for man (revised ed). RAND, Santa Monica, CA. https://www.rand.org/pubs/commercial_books/CB179-1.html
11. Jaime LG et al (2012) Regions of dynamical stability for discs and planets in binary stars of the solar neighbourhood. Mon Not Roy Astronom Soc 427(4):2723–2733. https://doi.org/10.1111/j.1365-2966.2012.21839.x

12. Howard AW, Marcy GW, Fischer DA et al (2014) The NASA-UC-UH Eta-Earth program. IV. A low-mass planet orbiting an M dwarf 3.6 pc from Earth. Astrophys J 794(1):51
13. Pinamonti M, Damasso M, Marzari F et al (2018) The HADES RV Programme with HARPS-N at TNG-VIII. GJ 15A: a multiple wide planetary system sculpted by binary interaction. Astronom Astrophys 617:A104
14. Teegarden BJ, Pravdo SH, Hicks M et al (2003) Discovery of a new nearby star. Astrophys J 589(1):L51
15. Zechmeister M, Dreizler S, Ribas I et al (2019) The CARMENES search for exoplanets around M dwarfs: two temperate Earth-mass planet candidates around Teegarden's Star. Astronom Astrophys 627:A49
16. Dreizler S, Luque R, Ribas I (2024) Teegarden's Star revisited: A nearby planetary system with at least three planets. Astronom Astrophys 684:A117
17. University of Puerto Rico at Arecibo (2024) Earth Similarity Index (ESI). https://phl.upr.edu/projects/earth-similarity-index-esi. Accessed 11 Jan 2024
18. Wandel A, Tal-Or L (2019) On the habitability of Teegarden's Star planets. Astrophys J Lett 880(2):L21
19. Wandel A, Gale J (2020) The bio-habitable zone and atmospheric properties for planets of red dwarfs. Int J Astrobiol 19(2):126–135. https://doi.org/10.1017/S1473550419000235
20. Ostberg C, Kane SR, Li Z et al (2023) The demographics of terrestrial planets in the Venus Zone. Astronom J 165(4):168
21. Pierrhumbert R (2010) Principles of planetary climate. Cambridge University Press, New York
22. Wolf ET, Toon OB (2015) The evolution of habitable climates under the brightening Sun. J Geophys Res Atmos 120(12):5775–5794
23. Del Genio AD, Way MJ, Amundsen DS et al (2019) Habitable climate scenarios for Proxima Centauri b with a dynamic ocean. Astrobiology 19(1):99–125
24. Miguel Y, Cridland A, Ormel CW et al (2020) Diverse outcomes of planet formation and composition around low-mass stars and brown dwarfs. Mon Not Roy Astronom Soc 491(2):1998–2009. https://doi.org/10.1093/mnras/stz3007
25. Dole SH (1970) Computer simulation of the formation of planetary systems. Icarus 13(3):494–508
26. Raymond SN, Quinn T, Lunine JI (2004) Making other earths: dynamical simulations of terrestrial planet formation and water delivery. Icarus 168(1):1–17
27. Walsh K (2005) Water world, glacier world, dust world. Analog Science Fiction and Fact, September 2005
28. Kopparapu RK, Ramirez R, Kasting JF et al (2013) Habitable zones around main-sequence stars: new estimates. Astrophys J 765(2):131
29. Gagné J, Faherty JK, Burgasser AJ et al (2017) SIMP J013656.5+093347 is likely a planetary-mass object in the Carina-Near Moving Group. Astrophys J 841(1):L1. https://doi.org/10.3847/2041-8213/aa70e2
30. Ibid.

31. Kao MM, Hallinan G, Pineda JS et al (2018) The strongest magnetic fields on the coolest brown dwarfs. Astrophys J Suppl Ser 237(2):25

32. Vos JM, Burningham B, Faherty JK et al (2023) Patchy forsterite clouds in the atmospheres of two highly variable exoplanet analogs. Astrophys J 944(2):138

33. Tan X, Showman AP (2021) Atmospheric circulation of brown dwarfs and directly imaged exoplanets driven by cloud radiative feedback: effects of rotation. Mon Not Roy Astronom Soc 502(1):678–699. https://doi.org/10.1093/mnras/stab060

34. L'Observatoire de Paris (2024) 82 Eridani. In: Encyclopaedia of exoplanetary systems. http://exoplanet.eu/catalog/. Accessed 4 Mar 2024

35. Basant R, Dietrich J, Apai D (2022) An integrative analysis of the rich planetary system of the nearby star e Eridani: ideal targets for exoplanet imaging and biosignature searches. Astronom J 164(1):12

36. Anderson P (1961) Orbit unlimited. Pyramid, New York

37. Anderson P (1982) New America. Tor, New York

38. University of Vienna (2024) Catalogue of exoplanets in binary star systems. https://adg.univie.ac.at/schwarz/multiple.html. Accessed 5 Mar 2024

39. Star wars: Episode IV—a new hope (1977) imdb.com. https://www.imdb.com/title/tt0076759/?ref_=nv_sr_srsg_4_tt_7_nm_0_q_star%2520wars. Accessed 6 Mar 2024

40. Feng F, Butler RP, Jones HRA et al (2021) Optimized modelling of Gaia–Hipparcos astrometry for the detection of the smallest cold Jupiter and confirmation of seven low-mass companions. Mon Not Roy Astronom Soc 507(2):2856–2868. https://doi.org/10.1093/mnras/stab2225

41. Mamajek EE, Hillenbrand LA (2008) Improved age estimation for solar-type dwarfs using activity-rotation diagnostics. Astrophys J 687(2):1264–1293. https://doi.org/10.1086/591785

42. Kuzuhara M, Fukui A, Livingston JH et al (2024) Gliese 12 b: A temperate Earth-sized planet at 12 pc ideal for atmospheric transmission spectroscopy. Astrophys J Lett 967(2):L21

43. Butler RP et al (1997) Three new 51 Pegasi-type planets. Astrophys J Lett 474(2):L115–L118. https://doi.org/10.1086/310444

44. Subbaraman N (2021) Elite US science academy expels astronomer Geoff Marcy following harassment complaints. Nature 592:159–160. https://www.nature.com/articles/d41586-021-01461-6

45. Malsky I, Rauscher E, Kempton EMR et al (2021) Modeling the high-resolution emission spectra of clear and cloudy nontransiting hot Jupiters. Astrophys J 923(1):62

46. Canup RM, Ward WR (2006) A common mass scaling for satellite systems of gaseous planets. Nature 441(7095):834–839. https://doi.org/10.1038/nature04860

47. McArthur BE, Benedict GF, Barnes R et al (2010) New observational constraints on the υ Andromedae system with data from the Hubble Space Telescope and

Hobby-Eberly Telescope. Astrophys J 715(2):1203. https://doi.org/10.1088/0004-637X/715/2/1203

48. Pizkorz D et al (2017) Detection of water vapor in the thermal spectrum of the non-transiting hot Jupiter Upsilon Andromedae b. Astronom J 154(2):78. https://doi.org/10.3847/1538-3881/aa7dd8
49. Ridpath I (2018) Star tales. Lutterworth, Cambridge, p 68
50. Ridpath I (2018) Star tales. Lutterworth, Cambridge, p 37
51. Ridpath I (2018) Star tales. Lutterworth, Cambridge, p 142
52. Frazer JG (ed) (1921) Apollodorus, the library. Harvard University Press, Cambridge, MA, 2.4.5. http://www.perseus.tufts.edu/hopper/text?doc=urn:cts:greekLit:tlg0548.tlg001.perseus-eng1:2.4.5
53. Clash of the Titans (1981) imdb.com. https://www.imdb.com/title/tt0082186/?ref_=fn_al_tt_2. Accessed 8 Mar 2024
54. Clash of the Titans (2010) imdb.com. https://www.imdb.com/title/tt0800320/?ref_=fn_al_tt_1. Accessed 8 Mar 2024
55. Smith EE (1950) Galactic patrol. Fantasy, Reading, PA
56. Selection from Galactic Patrol by E. E. "Doc" Smith, copyright © 1937, 1965 by the Literary Estate of E. E. "Doc" Smith; first appeared in Astounding Stories; used by permission of the Author's Estate and the Estate's Agents, the Virginia Kidd Agency, Inc.
57. Riley L (2024) The seven sisters series: myths and legends of the Pleiades. https://lucindariley.co.uk/myths-and-legends/. Accessed 9 Mar 2024
58. National Park Service (2022) First stories. Devil's Tower national monument, Wyoming. https://www.nps.gov/deto/learn/historyculture/first-stories.htm. Accessed Mar 9 2024
59. Close encounters of the third kind (1977) imdb.com. https://www.imdb.com/title/tt0075860/. Accessed 9 Mar 2024
60. Schiavi R, Capuzzo-Dolcetta R, Arca-Sedda M, Spera M (2020) Future merger of the Milky Way with the Andromeda galaxy and the fate of their supermassive black holes. Astronom Astrophys 642:A30. https://doi.org/10.1051/0004-6361/202038674
61. Fields BD, Melott AL, Ellis J et al (2020) Supernova triggers for end-Devonian extinctions. Proc Natl Acad Sci USA 117(35):21008–21010
62. Melott AL, Thomas BC (2011) Astrophysical ionizing radiation and Earth: a brief review and census of intermittent intense sources. Astrobiology 11(4):343–361
63. Thomas BC, Yelland AM (2023) Terrestrial effects of nearby supernovae: updated modeling. Astrophys J 950(1):41
64. Clarke AC (1958) The star. In: The other side of the sky. Harcourt Brace, San Diego, CA

8

The River of Stars
The Epsilon Eridani Sector: 0–6 h, 0 to –90°

The river that runs through this sector is the constellation Eridanus. The origin of its name is unclear [1]. The river is a crucial part of the myth of Phaeton, the son of Helios, the Sun-god. Phaeton stole his father's fiery chariot, careering across the sky and starting fires on Earth. Eventually, Zeus struck him down with a thunderbolt and he plunged into the Eridanus [2]. The origin of this story could be astronomical, possibly an ancient memory of the fall of a large meteorite. Exactly where the meteorite fell is hard to say, as many cultures throughout the world tell similar stories [3].

The star chosen to represent this sector is Epsilon Eridani, a name that will be familiar to readers of science fiction. This nearby K star is the site of many fictional planets. The space station *Babylon 5* might be located in this system. There appears to be some controversy about this, but it likely orbits Epsilon III, a lifeless, carbon-dioxide world [4]. In Asimov's *Foundation* series, the ice-cold planet Comporellon is located here, and despite its short summers, it hosts orbiting farm settlements that enable it to be a well-known exporter of pineapples [5]. In the *Revelation Space* series by Alastair Reynolds, the capital of the rather hostile planet Yellowstone, Chasm City, is located in a huge crater whose volcanic gases enable a breathable atmosphere to be provided for the domed city (Fig. 8.1).

In reality, the Epsilon Eridani system is very young and very dusty. The system hosts a very narrow Kuiper belt, at a distance of about 70 a.u. from its star, and either a broad asteroid belt from about 3 to about 20 a.u., or two asteroid belts, at 2 and 8 a.u [6]. In total, the belts contain many times the mass of our solar system's asteroid and Kuiper belts combined [7]. In the regions close to the habitable zone of Epsilon Eridani, this system has

© The Author(s), under exclusive license to Springer Nature Switzerland AG 2024
K. J. E. Walsh, *Planets of the Known Galaxy*, Science and Fiction,
https://doi.org/10.1007/978-3-031-68218-6_8

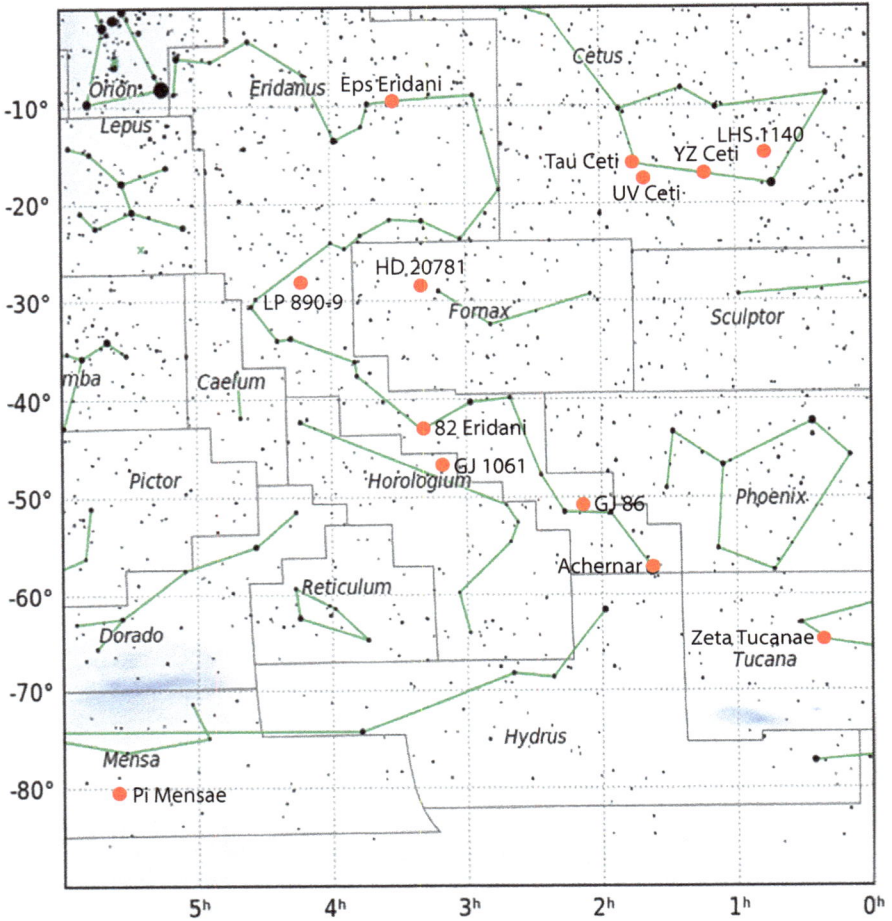

Fig. 8.1 Map of the Epsilon Eridani sector. Blue shading indicates the "Milky Way", the most star-packed part of the galaxy, while the blue (darker) line is the galactic equator, an imaginary circle running through the Milky Way that slices the galaxy in half. The orange (lighter) line is the path of the Sun's location through the seasons (indicating the "zodiac" constellations). Larger circles indicate brighter stars, while locations mentioned in this chapter are indicated by red circles. Image produced using StarCharter (https://github.com/dcf21/star-charter, accessed on the 2nd of July 2024. © 2020, Dominic Ford) software under the GNU general public license

hundreds of times the amount of dust that occurs in the same part of our solar system [8]. The source of this habitable zone dust could be comets, as it is in our solar system [9, 10]. If so, that might be bad news for the potential habitability of terrestrial planets in the Epsilon Eridani system, as it might mean more giant impacts from comets. The biggest asteroid or comet impact on

Earth in historical times was the Tunguska event in Siberia in 1908, estimated to have had an explosive power of about ten megatons, big enough to destroy a city. The frequency of similar events is difficult to estimate but might be about once every 1000 years. If in the Epsilon Eridani system there were (say) 100 times the number of impacts that occur in our system, this would mean a Tunguska-size event every 10 years, and an impact like the one that killed the dinosaurs would happen every few million years, instead of every few hundred million years as on Earth [11, 12]. This could seriously disrupt the evolution of life on habitable planets in that system.

The star Epsilon Eridani is now officially known as Ran, and wandering among the asteroid belts is the planet Ægir, a cold Jovian. In Norse myth, Ægir is the personification of the sea as a friend, but Ran, his wife, rules a realm at the bottom of the sea where drowned men go [13]. That must make for some interesting dinnertime conversation.

A giant planet like Ægir could host some large moons. We've already talked about how tidal heating can melt the interiors of the moons of giant planets, like the moon Io circling Jupiter. Recently, PhD student Elina Kleisioti and her co-workers at Leiden University have outlined how extreme tidal heating of moons of Ægir could lead to their discovery [14]. Tidal heating could warm the surface of the moon to the point where it could be detected by an infrared telescope. Their calculations show that a moon twice the radius of Io with reasonable values of eccentricity of its orbit around Ægir would have surface temperatures above freezing and thus would become detectable by the JWST.

Less than two parsecs from Ran is a G star with another famous name, Tau Ceti. There are so many places there to talk about. It is the location of Pell's World and Downbelow Station in Cherryh's Alliance-Union series [15]. Superficially similar to Earth, Pell's World is perpetually cloudy, apparently due to the high concentration of carbon dioxide in its atmosphere. Like Pandora, this is a world where humans require face masks to breathe. Meteorologically, though, it is not clear that a planet with a higher carbon dioxide concentration in its atmosphere would necessarily be cloudier. On Pell's World, the presence of the Hisa, sentient and intelligent local life forms, enormously complicates the human settlement of the planet. The Tau Ceti system is also the location of Ursula Le Guin's classic novel *The Dispossessed* (1974), set in the Hainish universe on the twin planets Urras and Anarres [16]. The beautiful, fertile world of Urras is home to a hyper-capitalist society, whereas the ugly, dry Anarres is run as a rational anarchy. A scientist on Anarres encounters obstacles when he tries to bring his work to fruition, so he decides to try his luck on Urras instead, with disastrous results.

But could twin planets like Urras and Anarres even exist? The jury is still out. Simulations indicate that twin planets could be created by a glancing impact of two objects of similar mass that formed in different locations early in the development of a stellar system. The two planets would have to crash into each other at specific speeds and angles so that they would end up circling each other instead of flying off into space, merging or just smashing each other to pieces [17, 18]. The result would be twin Earth-sized planets orbiting within perhaps 20,000–30,000 km of each other. Each planet would raise massive ocean tides on the other and would be awesomely enormous in the night sky [19]. There are also more complicated ways in which twin habitable planets could form, involving gravitational interactions with other bodies [20]. Since no twin planets have yet been found, it is very hard to say how common they really are. But based on what we know so far, twin planets are likely to be rare, and twin habitable planets even rarer.

In Larry Niven's *A Gift From Earth* (1968), part of the Known Space universe, the Tau Ceti system hosts the planet Plateau, a Venus-like world with a surface temperature of about 300 °C (about 600 °F) [21]. The only habitable region is on a coolish, barren plateau 60 km high, called Mount Lookitthat after what the captain of the first colony ship said upon sighting it (i.e. "Lookitthat!!"). Rivers end in great waterfalls that cascade off its edge. Its flat surface is gashed by cliffs and canyons, and Niven describes the terrain as "cruel to a mountain goat".

It is unknown whether such high topography could exist. The peak of Olympus Mons on Mars is about 21 km above the surrounding plains, and it has a large shield volcano region that is about the size of Poland, much of it higher than 10 km [22]. On Venus, the Maxwell Montes region is about 11 km above the mean planetary elevation. Neither of these elevations approach that of Plateau. On a planet like Earth, with a thin crust and a mantle that acts like a sticky fluid, very tall topography tends to sink into the crust, due to its own weight. One relevant factor here is that Tau Ceti has a much higher ratio of magnesium to silicon than the Sun does. Assuming that this ratio is reflected in the compositions of its planets, this would mean that they would have a less viscous mantle, and this appears to be associated with smaller variations in surface topography [23]. This makes sense, as a stickier mantle would be able to support higher topography. So, 60 km? Right now, it's fair to say that we probably do not know enough about planetary geology to answer this question.

The Tau Ceti system is also visited in Heinlein's classic juvenile novel *Time for the Stars* (1956) [24]. The eminently habitable planet Constance orbiting Tau Ceti appears welcoming at first to the crew of the starship *Lewis and*

Clark, but it hides a dark secret. Unlike many of Heinlein's other books for young adults, this novel has an unusually thoughtful treatment of interpersonal and psychological issues, in this case those associated with throwing a crew together and casting them off into interstellar space with the understanding that they will probably never see any of their friends and family back on Earth ever again. Less attractive is his characterization of female crewmembers as clearly the weaker sex. Equally unappealing is his acceptance of the need for humanity to colonize the universe by force, even in the teeth of what turns out to be organized, intelligent resistance from indigenous aliens of a planet in the Beta Hydri system, an old G star in this sector about seven parsecs away from Earth.

The Tau Ceti system is also where Barbarella crash-landed, supposedly on its 16th (!) planet, in the 1967 Jane Fonda film. No comment.

More recently, in Kim Stanley Robinson's *Aurora* (2015), colonists arrive to settle an Earth-like moon of the super-Earth Tau Ceti e [25]. This brilliantly written novel covers a lot of territory, from the biological and psychological effects of interstellar travel, to the future of humanity and its relationship with artificial intelligence, to the meaning of life. The novel has a particularly strong scientific backstory, with a lot of effort made to include realistic and up-to-date details of the planets of the Tau Ceti system. In the book, the moon of Tau Ceti e is slightly smaller than Earth and has a thinner atmosphere, but one that is still breathable. One important difference from Earth is that the moon's atmosphere is apparently not produced by biological processes, but instead is one of those worlds with substantial abiotic oxygen (but there is a plot twist).

There are a number of clever touches in the description of this moon, named Aurora. It is said to have an orbital period of about 18 days, and since it is tidally locked and synchronously rotating, this is also its day length. The moon is described as mostly oceanic and is very windy as a result. Combined with gravity less that Earth's, this creates enormous ocean waves. Despite the moon's slow rotation, it is described as having a strong magnetic field, good for retaining its atmosphere over the billions of years of its life to date. The super-Earth Tau Ceti e is almost certainly larger than four Earth masses, so as in *Aurora*, it could conceivably host a captured Earth-sized moon. In the novel, the moon is said to have a mass of 0.83 times that of Earth, giving a mass ratio between the moon and its planet of about 1:5 or so. This ratio is not out of the question, as it only a little larger than the mass ratio of Pluto to its moon Charon (about 1:7). The latest theory on the formation of the Pluto-Charon system is that they were two bodies who gently collided but then ended up orbiting each other, a little like the proposed formation mechanism for twin planets [26]. Thus an Earth-sized moon of Tau Ceti e could form that way.

This hypothetical moon could be hot, as its estimated equilibrium temperature would be 290 K, well above Earth's. If it had an atmosphere that was thinner than Earth's and with less greenhouse gases, that would help cool it. It would also help if the planet were dry, rather than wet as described in *Aurora*. Desert worlds likely have wider habitable zones than water worlds [27]. For desert worlds, runaway greenhouse is more difficult to achieve because the runaway mechanism relies on the greenhouse effect from water vapor, a molecule in shorter supply on dry worlds. For desert worlds, the inner edge of the habitable zone is thus closer to the star. The outer edge of the habitable zone is also further away from the star for dry worlds. They do not have much ice and so have a lower albedo, making them warmer. It is easier for them to avoid runaway glaciation, where a world cools, gains more ice, reflects more sunlight, and cools some more, eventually leading to a snowball world.

A slow planetary rotation would also help to protect against runaway greenhouse. As mentioned earlier, the day side of slowly rotating planets tends to be overcast, reflecting incoming solar radiation and cooling the planet [28]. In the case of Tau Ceti e, it receives about 70% more stellar radiation than Earth does. This is less than Venus, which these days has about 90% more but would have had less in the past when the Sun was not as bright. Simulations of the past climate of Venus are relevant here, as published in 2016 by NASA's Michael Way and collaborators [29]. About 700 million years ago, Venus received about 70% more radiation than Earth does today, or about the same as Tau Ceti e receives today. At that amount of radiation and assuming an Earth-like atmospheric composition, simulations show that a rotation period the same as modern-day Venus (i.e. hundreds of days) would have been enough to cool the planet to temperatures similar to those of present-day Earth. In this simulation, temperatures on the day side are well above freezing, up to 35 °C (95 °F) in substellar land regions, while on the night side, it is cold enough to snow. Despite the large day-night temperature differences, this is still largely a habitable climate for humans. If the sol is shortened to 16 Earth days, though, everything changes. The planet becomes essentially uninhabitable, with day side maxima of more than 80 °C and night side minima only cooling down to 27 °C (81 °F). In *Aurora*, Tau Ceti e has a sol of about 18 days, so this would not be that helpful in cooling the planet.

The next planet further out, Tau Ceti f, might be in the habitable zone, but it is close the outer edge. Moreover, due to the gradual increase of brightness of its star, this planet might only have been in the habitable zone for about 1 billion years [30]. That is a really short time for complex life to develop. In the

Table 8.1 Planets of the Tau Ceti system

Planet	Mass (Earth = 1)	Orbital period (days)	Distance from star (Earth distance from Sun = 1)	Equilibrium temperature (degrees K, assuming albedo = 0.3) (Earth = 255 K)	Type
Tau Ceti g	>1.74	20	0.133	583	Hot super-Earth or sub-Neptune
Tau Ceti h	>1.83	49.4	0.243	432	Hot super-Earth or sub-Neptune
Tau Ceti e	>3.93	163	0.538	290	Hot super-Earth or sub-Neptune
Tau Ceti f	>3.93	636	1.334	184	Warm or cold super-Earth or sub-Neptune

novel, its moon has a Mars-like climate and atmosphere, so if life developed there at all, it would probably be microscopic.

The orbital parameters of the Tau Ceti system are not yet known precisely. The rotational axis of the star is likely inclined at about 7°, or almost face-on [31]. On the other hand, the system contains a massive debris disk that is inclined at a different angle, about 35°. Calculations of orbital stability suggest that the planets could not share the same orbital orientation as the axis of the star, as then their orbits would become unstable in a few million years. Since the system is probably older than our own solar system, this implies a different inclination [32]. If it is assumed that the inclination of the planetary orbits is the same as that of the debris disk, orbital stability analysis with the DYNAMITE software predicts a super-Earth planet in the habitable zone, one that has not yet been observed [33]. If this inclination is correct, the true masses of the planets in Table 8.1 would be about 1.75 times their minimum masses. This would make Tau Ceti e about 6.8 Earth masses, or more like a sub-Neptune rather a super-Earth, and so would be a very different world.

Tau Ceti e   *How similar to Earth: Larger, hotter, possible sub-Neptune.*
Plausible planet: Strongly heated small gas giant of a kind not seen in our solar system.
Best case scenario: The planet itself is probably too large and hot to be habitable but could host a more hospitable large moon.

Worst case scenario: Moons that are rocky and hot, with no atmosphere, or no moons at all.

Cultural connection: Aurora.

*   *   *

A short trip of less than a parsec from Tau Ceti is the Gliese 65 system, also known as Luyten 726-8. It consists of two M stars, one of which also has the designation UV Ceti, and it is one of the most extreme examples of a flare star known. It lends its name to a whole class of small M stars with intense flares. The flares of these UV Ceti stars rapidly reach a maximum brightness in seconds and last for about ten to fifty minutes. Increases in brightness range from about double their normal brightness to over 100 times as bright [34]. Also worth a quick visit is YZ Ceti, another UV Ceti type star only 0.5 parsecs away from Tau Ceti (Table 8.2). Its chief claim to fame is that one its planets, planet b, a hot world about the size of Earth or a little smaller, is emitting periodic bursts of radio waves. This is likely caused by the planet's magnetic field [35]. A planet orbiting an M star needs a strong magnetic field to stop its atmosphere from being stripped, and the fact that at least one planet in the YZ Ceti system may have such a magnetic field is reassuring for the potential habitability of other M star planets. The strength of the planetary magnetic field required to power the observed bursts is currently uncertain, but it could be stronger than Earth's. A lot more work is needed to characterize the environment of these planets, but in any event all of them are likely too hot to be habitable.

The M star GJ 1061 is in the constellation Horologium, the pendulum clock, only about 3.7 parsecs away from Earth. As one can deduce from its name, this is not an ancient constellation. It was created in the eighteenth century by Lacaille, who during his 2-year stint at Cape Province in South

Table 8.2 Planets of YZ Ceti

Planet	Mass (Earth = 1)	Orbital period (days)	Distance from star (Earth distance from Sun = 1)	Equilibrium temperature (degrees K) (Earth = 255 K)	Type
YZ Ceti b	>0.7	2.0	0.016	431–471	Hot rocky or Earth-size
YZ Ceti c	>1.14	3.1	0.022	375–410	Hot Earth-size or super-Earth
YZ Ceti d	>1.09	4.7	0.029	326–357	Hot Earth-size or super-Earth

Africa spent almost every night making observations of thousands of Southern Hemisphere stars [36]. After he finished compiling his catalogue, he was stuck in Cape Town for a while due to contrary winds for sailing. He then needed another task to occupy his time, so he made a calculation of Earth's radius. His result was out by about 0.2% from previous Northern Hemisphere calculations, a big discrepancy and one that worried him. It was decades before anyone figured out the source of the error. To measure latitude, Lacaille needed a plumb bob to give a vertical axis, but he did not account for the gravitational attraction on the bob of the nearby Table Mountain, which pulled the bob slightly away from the vertical.

There are three known planets in the GJ 1061 system (Table 8.3). While all are somewhat larger than Earth, and the inner two are likely too hot, planet d could have a reasonable climate, if a little on the cold side. If it is largely ice-covered, then one intriguing possibility is life developing under the ice [37]. Work by Lujendra Ojha of Rutgers and his team suggests that provided planet d has an Earth-like composition, if it is ice-covered the heat from its interior should be enough to melt the base of the ice sheet to create a sub-glacial ocean, a potential habitat for life. As the metal content of the star is very similar to that of our Sun, there should be enough long-lived radioactive elements like uranium and thorium in the planet's interior to maintain internal heat. The planet is also larger than Earth, and all things being equal, a larger planet should have a larger amount of interior heat per square meter of surface area. This potentially life-bearing ocean under the ice would be similar to the seas known to exist under the icy surfaces of Jupiter's moon Europa and Saturn's moon Enceladus. The main deal breaker for habitability is, as usual, the possible lack of a strong magnetic field. SEPHI calculates a low magnetic field strength for this likely slow-rotating, tidally locked world, and this might make it difficult for it to retain an atmosphere. An atmosphere is needed because at its calculated effective temperature, only slightly less than Earth's, water ice would gradually evaporate in direct sunlight with no atmosphere.

Table 8.3 Planets of the GJ 1061 system

Planet	Mass (Earth = 1)	Orbital period (days)	Distance from star (Earth distance from Sun = 1)	Equilibrium temperature (degrees K) (Earth = 255 K)	Type
GJ 1061 b	>1.44	3.2	0.021	354–387	Hot Earth-size or super-Earth
GJ 1061 c	>1.74	6.7	0.035	274–299	Hot super-Earth
GJ 1061 d	>1.57	13.0	0.052	225–246	Warm Earth-size or super-Earth

On the other hand, the star is inactive and so probably is not doing much to strip the planet's atmosphere at this stage [38]. Like many such planets, all three of the planets in this system may be strongly tidally heated [39]. But this depends upon their eccentricities, which are currently not known precisely. The main point here is that large exoplanets with ice-covered oceans could be important locations for life. Whether we will ever be able to detect life on these worlds is not clear.

GJ 1061 d *How similar to Earth: Larger, likely colder.*
Plausible planet: Ice-covered ocean planet.
Best case scenario: Would likely require a largish greenhouse effect to have habit-
able temperatures.
Worst case scenario: Frozen solid, ocean under the ice.
Cultural connection: Life beneath the frozen sea?

*   *   *

The F9.5 star Zeta Tucanae is considered one of the top three candidates within the known galaxy that could host a habitable world, according to one assessment (the other two are Beta Canum Venaticorum and 61 Virginis) [40]. No planets have yet been detected orbiting it, though. One point to note: if planetary orbits are practically face on, then neither radial velocity nor transit observations will be able to detect planets. Astrometry will work, particularly for larger planets distant from their star. Direct imaging will also work, but with current technology only for very large worlds. There appears to be no information on the orientation of Zeta Tucanae's rotational axis, and this is important as planetary orbits often have the same or similar inclination (but not always, particularly for large planets orbiting close to hotter stars) [41]. A simple explanation for the lack of detection of planets around this very suitable F star might be that it actually has no planets. At present, we do not know how many stellar systems have no planets at all, because any hypothetical alien astronomers working in nearby stellar systems and using our current detection methods would not be able to find Uranus and Neptune in our own solar system. Actually, space-based microlensing could, but only purely by chance, if an observer just happened to be looking at the right place at the right time [42]. The huge drawback of the microlensing method is that the random drift of objects in front of the star that causes a planet to be detected usually cannot be repeated. For many of the stellar systems of the known galaxy, discovery still awaits.

Further Afield

The stellar systems in this sector that are more than 10 parsecs away challenge our understanding of planetary formation mechanisms. The planet GJ 86 b is a hot super-Jovian that has formed in a stellar system with a nearby white dwarf only about 20 a.u. away and on an eccentric orbit. When both stars were on the main sequence, the white dwarf would have been even closer to its companion star because at that time the white dwarf would have had more mass, as stars throw off gas during their red giant stage. Yet the super-Jovian planet formed anyway [43]. This implies that planetary formation can take place in more hostile environments than previously thought.

Another problematic system is that of the K star HD 20781, about 36 parsecs away. It is very close to the G star HD 20782, at least 9000 a.u. away, and both stars host planets. This was the first known example of both stars in a binary system with their own planets (as of 2024, there are two such systems known) [44]. The K star hosts yet another of those compact stellar systems where planets both big and small are crammed inside half an astronomical unit, and all four of them are too hot to be habitable [45].

The Jovian planet orbiting the G star is remarkable in that its orbit has the largest eccentricity known [46]. Such an orbit probably arose from gravitational interactions with the distant K star [45]. In any event, the Jovian has certainly ejected any planets that formed in its system near its own orbit. At closest approach, it is only 0.06 a.u. from its star, and at that distance it is being inundated by more than 300 times the amount of solar radiation that Earth experiences. At furthest distance, it receives less radiation than Mars does. Its eccentricity is comparable to that of Halley's comet, and because the planet is so strongly heated at closest approach, it might lose gas like a comet as well. An idea of the effect of this extreme heating on the planet's weather can be obtained from observations of another planet with a comet-like orbit, the super-Jovian HD 80606 b, located in the Procyon sector about 66 parsecs away. It approaches to about 0.03 a.u. from its G star, and during its rapid, fiery passage it heats up by several hundred degrees C in just a few hours, reaching as much as 1500 K [47]. This enormous heating causes an equally enormous response in the planet's atmosphere. Jet stream winds in the upper atmosphere race away from the heated region and are estimated to ramp up to more than 3000 km/h [48, 49]. Shock waves blast around the planet.

The weather is extraordinarily awful, but how is the eccentricity of the planet's orbit maintained against tidal dissipation? The star is old, so either there must be some mechanism keeping the orbit eccentric, or tidal dissipation must be weak, or some combination of the two. It is thought that the Kozai mechanism is at least partially responsible for the planet's orbit [50]. In Jovian planets with large atmospheres, it is also thought that tidal dissipation is mostly caused by internal waves in the atmosphere that soak up energy, in much the same way that ocean waves on Earth soak up energy from the wind. Internal wave dissipation is much more effective in rapidly rotating planets [51], but the rotation period of HD 80606 b is thought to be about 90 h or so, much slower than Jupiter. Indeed, calculations of the tidal dissipation rate for the super-Jovian give values a lot less than for Jupiter in our solar system [48].

Another system where any habitable planets must have been thrown out (if they ever existed) is the Pi Mensae system, a G star about 18 parsecs away with two or perhaps three surviving planets. Planet b is a super-Jovian that is almost large enough to be a brown dwarf [52]. On an eccentric orbit, it travels from about 5.4 a.u. away from its star, about the distance of Jupiter from the Sun, to about 1.2 a.u. away, or close to the orbit that an Earth-like planet might have in that system. Far too close for comfort. As a result, Pi Mensae is no longer a priority target in the search for Earth-like worlds.

In this system, there is also the transiting very hot super-Earth or sub-Neptune planet c, which likely has an atmosphere [53]. Antonio Garcia Muñoz of the Technical University of Berlin and his collaborators have analyzed Hubble Space Telescope observations of this world and found a spectral signature for calcium. The planet has a density about half that of Earth, while observations and modeling are most consistent with a thick atmosphere containing at least 50% "heavy volatiles" like water vapor and carbon dioxide. The planet's radius of about twice that of Earth is close to the dividing line separating mostly rocky planets like Earth from planets like Uranus that have a substantial gas portion. This leads to the intriguing possibility that pi Mensae c once had a thicker atmosphere but has been gradually losing it due to the super-abundance of stellar radiation that it currently "enjoys" (more than 300 times what Earth receives). Thus the planet might now be transitioning from a gassy sub-Neptune to a more rocky super-Earth.

Another peculiar aspect of this system is the strange inclination of its planetary orbits. Planet c transits, so at 88° its inclination is just about edge-on, but the orbit of planet c is inclined at 50°. It has been pointed out that explaining how this might have occurred provides important constraints for theories of planetary formation and for the stability of planetary orbits [54]. Another system with planets that have different orbital inclinations is Upsilon Andromedae (Chap. 7), so that system provides another test case.

A more conventional system is that of the M star LHS 1140, about 15 parsecs away in the constellation Cetus. The abbreviation "LHS" stands for "Luyten Half-Second Catalogue", a work compiled by Jacob Luyten of Luyten's Star fame, a catalogue of stars with very high proper motions [55]. Of most interest is planet b. Early measurements of its mass and radius indicated that it might be a rocky world, but a revised analysis of the observations by Charles Cadieux of the Université de Montréal published in 2024 gives a lower mass and density. This means that it is either a sub-Neptune or a waterworld, with water comprising a large proportion of its mass [56]. If it is a sub-Neptune, it is unlikely to be habitable, but the waterworld case is definitely of interest. This planet is one of the easiest to observe astronomically and thus is a high-priority target for further observations. With an equilibrium temperature of about 225 K, this warm super-Earth is also a top candidate for habitability studies. It will come as no surprise to readers that climate model simulations of the waterworld case give a wide range of predictions, depending on the assumptions made about the content of its atmosphere [57]. These simulations assume a global ocean, as given the planet's density it is unlikely that there would be any land surface. Synchronous rotation is assumed, as its orbit has low eccentricity, making it difficult to break out of synchronous rotation. It has also been shown that a planet with a thick atmosphere might be able to avoid synchronous rotation due to the presence of strong atmospheric tides, but in this case the planet's gravitational tides are likely too strong for its atmospheric tides to be able to stop the planet becoming tidally locked [58, 59].

For an Earth-like atmosphere, the simulations show that only the region at the sub-stellar point remains unfrozen. But for a 5-bar carbon dioxide atmosphere (i.e. five times the sea level atmospheric pressure on Earth, but entirely CO_2), almost all of the day side of the planet is ice-free, with temperatures at the sub-stellar point of about 30 °C (86 °F). The nifty thing about these climate model simulations is that they also make specific predictions that can be confirmed by astronomical observations. This planet transits and at some point the light from its star passes through its atmosphere, creating a so-called transmission spectrum. The output of the climate model can be used to calculate synthetic spectra, and these can then be compared with actual measured transmission spectra to help determine atmospheric composition. Cadieux and co-authors figure that about 20 transit measurements with the JWST would give enough data to decide whether the planet has a thinner secondary atmosphere rather than a thicker primordial atmosphere, and so whether is a waterworld or not. One problem is that because the planet is relatively far from its star, only about four transits per year can be measured using the

JWST, so several years' worth of measurements will be needed to verify the presence of a secondary atmosphere. In contrast, a thick primordial atmosphere could be confirmed in only one observation.

An important theoretical climate threshold is the amount of incoming radiation required to initiate runaway greenhouse. Along with Teegarden's Star b (Chap. 7), planet c of the system of the M star LP 890-9 is an excellent test of this concept, as this super-Earth lies close to the inner edge of the conservative habitable zone [60]. Its star has another designation, SPECULOOS-2, actually named after a European observing project, the Search for habitable Planets EClipsing ULtra-cOOl Stars, consisting of two large telescopes, one in Chile and the other in Tenerife—it is surely just a coincidence that Speculoos is also the name of a Belgian gingerbread biscuit [61].

Epsilon Eridani Sector Summary

The Sun-like star Tau Ceti beckons like a siren in this sector, its many fictional worlds almost begging to be matched by a real habitable planet. Pi Mensae and Epsilon Eridani have probably already failed this test, but Zeta Tucanae is still in the running. There are a number of systems in this sector whose characteristics challenge our understanding of planetary formation mechanisms. Finally, Eridanus meanders through this sector, flowing southwards until it reaches the bright star Achernar, actually a pair of main sequence B and A stars about 43 parsecs away. The name of the star is Arabic, meaning, naturally enough, "end of the river". Earthly rivers run into the sea, while Eridanus runs into the vast emptiness of interstellar space.

References

1. Barentine JC (2016) Phaeton. In: Uncharted constellations. Springer, Cham, pp 101–108. https://doi.org/10.1007/978-3-319-27619-9_11
2. Kline AS (trans) (2000) Ovid, metamorphoses. Book II. https://www.poetryin-translation.com/PITBR/Latin/Metamorph2.php
3. James P, van der Sluijs MA (2016) The fall of Phaethon in context: a new synthesis of mythological, archaeological and geological evidence. J Ancient Near East Relig 16(1):67–94. https://doi.org/10.1163/15692124-12341279
4. The Babylon 5 Project (2024) Epsilon III. https://babylon5.fandom.com/wiki/Epsilon_III. Accessed 14 Mar 2024
5. Asimov I (1986) Foundation and Earth. Doubleday, New York

6. Booth M, Pearce TD, Krivov AV et al (2023) The clumpy structure of ε Eridani's debris disc revisited by ALMA. Mon Not Roy Astronom Soc 521(4):6180–6194. https://doi.org/10.1093/mnras/stad938

7. Greaves JS, Wyatt MC, Holland WS, Dent WRF (2004) The debris disc around Tau Ceti: a massive analogue to the Kuiper Belt. Mon Not Roy Astronom Soc 351(3):L54–L58

8. Ertel S, Defrère D, Hinz P et al (2020) The HOSTS survey for exozodiacal dust: observational results from the complete survey. Astronom J 159(4):177

9. Kral Q, Krivov AV, Defrère D et al (2017) Exozodiacal clouds: hot and warm dust around main sequence stars. Astronom Rev 13(2):69–111

10. Su KY, De Buizer JM, Rieke GH et al (2017) The inner 25 au debris distribution in the ε Eri system. Astronom J 153(5):226

11. Mignan A, Grossi P, Muir-Wood R (2011) Risk assessment of Tunguska-type airbursts. Nat Hazard 56:869–880. https://doi.org/10.1007/s11069-010-9597-3

12. Nesvorný D, Bottke WF, Marchi S (2021) Dark primitive asteroids account for a large share of K/Pg-scale impacts on the Earth. Icarus 368:114621

13. Simek R, Hall A (trans) (1996) Dictionary of northern mythology. Boydell and Brewer, Martelsham

14. Kleisioti E, Dirkx D, Rovira-Navarro M, Kenworthy MA (2023) Tidally heated exomoons around ε Eridani b: Observability and prospects for characterization. Astronom Astrophys 675:A57

15. Cherryh CJ (1981) Downbelow station. DAW, New York

16. Le Guin UK (1974) The dispossessed. Harper and Row, New York

17. Ryan K, Nakajima M, Stevenson DJ (2014) Binary planets. In: Abstracts of the American Astronomical Society, DPS meeting #46, Tucson, 9–14 Nov 2014. https://ui.adsabs.harvard.edu/abs/2014DPS....4620102R/abstract

18. California Institute of Technology (2014) Can binary terrestrial planets exist? phys.org. https://phys.org/news/2014-12-binary-terrestrial-planets.html

19. Walsh K (2023) Urras, Anarres and Marsmoon. Analog Science Fiction and Fact, July/August 2023

20. Chrenko O, Brož M, Nesvorný D (2018) Binary planet formation by gas-assisted encounters of planetary embryos. Astrophys J 868(2):145

21. Niven L (1968) A gift from Earth. Ballantine, New York

22. Wikipedia (2024) Olympus Mons. Accessed 13 Mar 2024

23. Guimond CM, Rudge JF, Shorttle O (2022) Blue marble, stagnant lid: could dynamic topography avert a waterworld? Planet Sci J 3(3):66

24. Heinlein RA (1956) Time for the stars. Ballantine, New York

25. Robinson KS (2015) Aurora. Orbit, London

26. McKinnon WB, Stern SA, Weaver HA et al (2017) Origin of the Pluto–Charon system: constraints from the New Horizons flyby. Icarus 287:2–11

27. Abe Y, Abe-Ouchi A, Sleep NH, Zahnle KJ (2011) Habitable zone limits for dry planets. Astrobiology 11(5):443–460

28. Way MJ, Del Genio AD, Kiang NY et al (2016) Was Venus the first habitable world of our solar system? Geophys Res Lett 43(16):8376–8383
29. Ibid.
30. Pagano M, Truitt A, Young PA, Shim SH (2015) The chemical composition of τ Ceti and possible effects on terrestrial planets. Astrophys J 803(2):90
31. Korolik M, Roettenbacher RM, Fischer DA et al (2023) Refining the stellar parameters of τ Ceti: a pole-on solar analog. Astronom J 166(3):123
32. L'Observatoire de Paris (2024) Tau Ceti. In: Encyclopaedia of exoplanetary systems. http://exoplanet.eu/catalog/. Accessed 14 Mar 2024
33. Dietrich J, Apai D (2020) An integrated analysis with predictions on the architecture of the τ Ceti planetary system, including a habitable zone planet. Astronom J 161(1):17
34. Gershberg RE, Katsova MM, Lovkaya MN et al (1999) Catalogue and bibliography of the UV Cet-type flare stars and related objects in the solar vicinity. Astronom Astrophys Suppl Ser 139(3):555–558
35. Pineda JS, Villadsen J (2023) Coherent radio bursts from known M-dwarf planet-host YZ Ceti. Nat Astron 7:569–578. https://doi.org/10.1038/s41550-023-01914-0
36. Warner B (2002) Lacaille 250 years on. Astronom Geophys 43(2):2.25–2.26. https://doi.org/10.1046/j.1468-4004.2002.43225.x
37. Ojha L, Troncone B, Buffo J et al (2022) Liquid water on cold exo-Earths via basal melting of ice sheets. Nat Commun 13:7521. https://doi.org/10.1038/s41467-022-35187-4
38. Dreizler S, Jeffers SV, Rodríguez E et al (2020) RedDots: a temperate 1.5 Earth-mass planet candidate in a compact multiterrestrial planet system around GJ 1061. Mon Not Roy Astronom Soc 493(1):536–550. https://doi.org/10.1093/mnras/staa248
39. McIntyre SRN (2022) Tidally driven tectonic activity as a parameter in exoplanet habitability. Astronom Astrophys 662:A15. https://doi.org/10.1051/0004-6361/202141112
40. Porto de Mello G, Fernandez del Peloso E, Ghezzi L (2006) Astrobiologically interesting stars within 10 parsecs of the Sun. Astrobiology 6(2):308–331
41. Winn JN, Fabrycky DC (2015) The occurrence and architecture of exoplanetary systems. Annu Rev Astronom Astrophys 53:409–447
42. Zhu W, Dong S (2021) Exoplanet statistics and theoretical implications. Annu Rev Astronom Astrophys 59:291–336
43. Zeng Y, Brandt TD, Li G et al (2022) The Gliese 86 binary system: a warm Jupiter formed in a disk truncated at ~ 2 au. Astronom J 164(5):188
44. L'Observatoire de Paris (2024) Planets in binary systems. In: Encyclopaedia of exoplanetary systems. http://exoplanet.eu/catalog/. Accessed 17 Mar 2024
45. Udry S, Dumusque X, Lovis C et al (2019) The HARPS search for southern extra-solar planets. XLIV. Eight HARPS multi-planet systems hosting 20 super-Earth and Neptune-mass companions. Astronom Astrophys 622:A37. https://doi.org/10.1051/0004-6361/201731173

46. O'Toole SJ, Tinney CG, Jones HRA et al (2009) Selection functions in Doppler planet searches. Mon Not Roy Astronom Soc 392(2):641–654. https://doi.org/10.1111/j.1365-2966.2008.14051.x

47. Laughlin G et al (2009) Rapid heating of the atmosphere of an extrasolar planet. Nature 457(7229):562–564. https://doi.org/10.1038/nature07649

48. Tsai S-M, Steinrueck M, Parmentier V et al (2023) The climate and compositional variation of the highly eccentric planet HD 80606 b—the rise and fall of carbon monoxide and elemental sulfur. Mon Not Roy Astronom Soc 520(3):3867–3886. https://doi.org/10.1093/mnras/stad214

49. Langton J, Laughlin G (2008) Hydrodynamic simulations of unevenly irradiated Jovian planets. Astrophys J 674(2):1106

50. Wu Y, Murray N (2003) Planet migration and binary companions: The case of HD 80606 b. Astrophys J 589(1):605

51. Lazovik YA, Barker AJ, de Vries NB, Astoul A (2024) Tidal dissipation in rotating and evolving giant planets with application to exoplanet systems. Mon Not Roy Astronom Soc 527(3):8245–8256. https://doi.org/10.1093/mnras/stad3689

52. Jones HR, Butler RP, Tinney CG et al (2002) A probable planetary companion to HD 39091 from the Anglo-Australian Planet Search. Mon Not Roy Astronom Soc 333(4):871–875. https://doi.org/10.1046/j.1365-8711.2002.05459.x

53. Muñoz AG, Fossati L, Youngblood A et al (2021) A heavy molecular weight atmosphere for the super-Earth π Men c. Astrophys J Lett 907(2):L36

54. Hatzes AP, Gandolfi D, Korth J et al (2022) A radial velocity study of the planetary system of π Mensae: improved planet parameters for π Mensae c and a third planet on a 125 day orbit. Astronom J 163(5):223

55. Luyten WJ (1979) Catalogue of stars with proper motions exceeding 0″5 annually. University of Minnesota, Minneapolis, MN

56. Cadieux C, Plotnykov M, Doyon R (2024) New mass and radius constraints on the LHS 1140 planets: LHS 1140 b is either a temperate mini-Neptune or a water world. Astrophys J Lett 960(1):L3

57. Ibid.

58. Ribas I, Bolmont E, Selsis F et al (2016) The habitability of Proxima Centauri b-I. irradiation, rotation and volatile inventory from formation to the present. Astronom Astrophys 596:A111

59. Leconte J, Wu H, Menou K, Murray N (2015) Asynchronous rotation of Earth-mass planets in the habitable zone of lower-mass stars. Science 347(6222):632–635

60. Quirino Q, Gilli G, Kaltenegger L et al (2023) 3D Global climate model of an exo-Venus: a modern Venus-like atmosphere for the nearby super-Earth LP 890-9 c. Mon Not Roy Astronom Soc Lett 523(1):L86–L91. https://doi.org/10.1093/mnrasl/slad045

61. ESO (2018) First light for SPECULOOS. https://www.eso.org/public/news/eso1839/

9

The Legacy of Lacaille
The Fomalhaut Sector: 18–24 h, 0 to –90°

This sector ranges from star-packed regions towards the center of the galaxy in Sagittarius to sparsely-starred constellations with unfamiliar names like Telescopium and Microscopium. The brightest star in this sector is Fomalhaut, an A star a little less than eight parsecs away, residing in the unexceptional constellation Piscis Austrinus ("the southern fish"). Fomalhaut is only about 400 million years old, so it is probably too young for much life to have developed on any of its planets [1]. Its system is also extremely dusty, so dusty that a large blob of dust orbiting the star was initially taken to be a planet, Fomalhaut b. Later observations showed that this blob was becoming both dimmer and bigger, which is very un-planet-like behavior. The blob is probably debris from a collision between some planetary precursor objects, or planetesimals as they are called [2], a collision that might have happened as recently as 2004 [3]. Such collisions in dynamically stable systems should be very rare, so it is possible that planets in the Fomalhaut system are currently migrating about before they settle down into permanent orbits (Fig. 9.1).

One of the most moving tales set in this system is *The Dowry of Angyar* (1964), Ursula Le Guin's famous short story of legend and heartbreak [4]. Later incorporated as the prologue to her 1966 novel *Rocannon's World*, it takes place in the Hainish universe, where (for a change) human beings are not the best and brightest, but just one of many races learning how to coexist and explore the stars [5]. Rocannon is a scientist and ethnologist on a mission to contact and study the inhabitants of Fomalhaut II, a very Earth-like planet with days of 29 standard hours but years of 800 standard Earth days. His mission goes very badly very quickly. This heroic fantasy/science fiction novel

© The Author(s), under exclusive license to Springer Nature Switzerland AG 2024
K. J. E. Walsh, *Planets of the Known Galaxy*, Science and Fiction,
https://doi.org/10.1007/978-3-031-68218-6_9

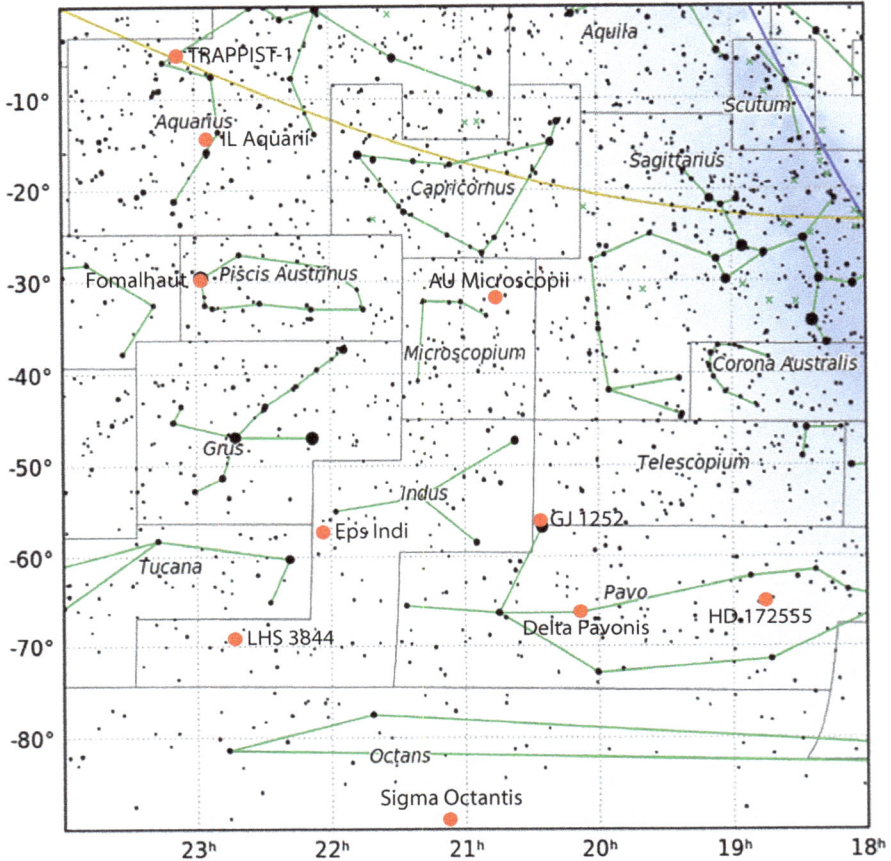

Fig. 9.1 Map of the Fomalhaut sector. Blue shading indicates the "Milky Way", the most star-packed part of the galaxy, while the blue (darker) line is the galactic equator, an imaginary circle running through the Milky Way that slices the galaxy in half. The orange (lighter) line is the path of the Sun's location through the seasons (indicating the "zodiac" constellations). Larger circles indicate brighter stars, while locations mentioned in this chapter are indicated by red circles. Image produced using StarCharter (https://github.com/dcf21/star-charter, accessed on the 2nd of July 2024. © 2020, Dominic Ford) software under the GNU general public license

touches on themes of how history can become legend, how science can interact with the occult and perhaps learn from it, and how a brave man deals with permanent exile.

Unusually, Fomalhaut has a couple of very distant companions that are also thought to be part of its stellar system. At 0.28 parsecs or about 57,000 a.u. away from Fomalhaut is the K star TW Piscis Austrini, sometimes called Fomalhaut B. The same age as Fomalhaut A, this K star is also perhaps too young to have a planet with complex life, but certainly not too young to have fully-formed planets, although none have been found yet. From a hypothetical planet in the system of Fomalhaut B, Fomalhaut A would be a dazzling star about four times as bright as Venus seen from Earth. Even further away is the M star Fomalhaut C, at 0.77 parsecs or 160,000 a.u. from Fomalhaut A. It is so far away from the main star that it is in a different constellation, Aquarius.

The zodiacal constellation Aquarius is arguably one of the most famous constellations. Its name means "water-carrier" in Latin, and many mythologies associate the constellation with this concept [6]. Some readers of a certain age will remember the hype in the 1960s associated with the "Age of Aquarius", whereby the world was about to enter a new era of harmony and understanding. Probably we are still waiting.

Aquarius has no genuinely bright stars but it does have a lot of planets. The M star Gliese 876, otherwise known as IL Aquarii, is a variable star of the BY Draconis type. Its planetary system is remarkable for a few reasons (Table 9.1). The orbital periods of the outer three planets are multiples of each other, almost exactly in the ratio 1:2:4, a resonant orbital pattern like that of the six planets of HD 110067 (Chap. 4). The masses of the planets of the IL Aquarii system are known accurately because they have been determined using a combination of radial velocity and astrometry measurements [7]. The orbits of all the planets are slightly eccentric, and the strong resonance is causing the direction of their closest approach to their star to change at the staggering rate of about 40° per year, thus taking only 9 years to move around a complete circle.

The innermost planet, planet d, is not in orbital resonance with the others, but its slightly eccentric orbit has strong implications for its environment. The planet has the potential for enormous amounts of tidal heating and may be entirely molten [8]. Even though the planet receives about 30 times as much

Table 9.1 Planetary system of IL Aquarii (Gliese 876)

Planet	Mass (Earth = 1)	Orbital period (days)	Distance from star (Earth distance from Sun = 1)	Equilibrium temperature (degrees K) (Earth = 255 K)	Type
IL Aquarii d	6.7	1.9	0.02	594–650	Hot super-Earth or sub-Neptune
IL Aquarii c	234	30.1	0.13	238–260	Warm Jovian
IL Aquarii b	749	61.1	0.21	188–206	Warm super-Jovian
IL Aquarii e	16	123.6	0.34	149–163	Cold sub-Jovian

stellar radiation as Earth does, the amount of heat generated by tidal processes could very well be considerably more than that, and vastly more than occurs on the already highly volcanic moon of Jupiter, Io (see Introduction). Thus planet d may well be an enormous lava world.

This leads to the idea that there may be worlds where the tidal heating at the surface considerably exceeds the heating from the incoming solar radiation, meaning that the surface temperature of the planet is largely determined by tidal effects. A planet like this could be a "tidal Venus", where tidal heating has been so large as to push an otherwise potentially habitable world into runaway greenhouse. Tidal Venuses were first described in a 2013 paper by Rory Barnes of the University of Washington and his collaborators [9]. They concluded that planets orbiting stars of less than about 0.3 times as massive as the Sun were vulnerable to this effect, or stars of about class M3 or smaller. An important point is that even though the orbital eccentricity that drives tidal heating may have occurred early in the evolution of a planet and would have been relatively short-lived, for some planets it would have lasted long enough to drive them into runaway greenhouse. There they would have stayed, even though the planet today might not experience much tidal heating because tidal dissipation has circularized its orbit.

The IL Aquarii system is also unusual in that it has two large planets in the conservative habitable zone. The Jovian planet IL Aquarii c could host an exomoon, as calculations suggest that gravitational and tidal effects, while significant, are not strong enough to cause such a moon to escape its orbit or to fall into its planet [10]. Planet b, a super-Jovian, is also in the habitable zone but further out. Tidal effects there from the star are less pronounced, so it also might be able to host a large exomoon.

While the age of IL Aquarii is uncertain, it is unlikely to be very young [11]. Some M stars, though, are young enough that their stellar systems are still forming. An example is the very young, largish M star AU Microscopii, about 9.7 parsecs away in the constellation Microscopium. This star is only about 20 million years old but has two planets and a very large debris disk [12]. The innermost one, planet b, is a recently discovered hot sub-Jovian and has been closely studied. The extreme youth of its stellar system provides an insight into how such systems develop. It is very likely that this planet is still losing mass, as it is hot enough, and since it has a low density, it is gassy enough as well. The mechanism for gas loss in this case appears to be heating of the upper layers of its atmosphere that accelerates some gas molecules to escape velocity, a process known as hydrodynamic escape. Recent observations and analysis by Dartmouth PhD student Keighley Rockliffe and collaborators suggest that the planet is indeed losing hydrogen [13]. This phenomenon could explain what is known as the "hot Neptune desert", the apparent shortage of this type of planet very close to stars [14]. A possible explanation is that gassy, Neptune-sized worlds do end up close to their stars, but then over millions of years they lose a lot of their gas, eventually becoming super-Earths or sub-Neptunes.

In contrast, planet c in this system, also a hot sub-Jovian, has a density more than three times higher than that of planet b. This suggests a planet with some rock, a large water component, but considerably less gas than planet b [12]. Being a very young M star, AU Microscopii is very flarey, so current estimated atmospheric mass loss from both planets b and c could be a significant portion of their atmospheres [15]. There might also be a planet d, of more Earth-like mass [16]. If so, this would be the first Earth-size planet found orbiting a very young star. It may transit, but this has not yet been detected. If it does transit, it would be a high-priority observation target, in order to examine the early evolution of the atmosphere of a terrestrial planet orbiting an M star.

South of AU Microscopii is a name familiar to readers of science fiction, Epsilon Indi, a multi-star system only 3.6 parsecs away. It is the possible home system of the Andorians, blue-skinned aliens who live on a frozen planet, and who appear in several episodes of the various *Star Trek* series. The main star of the system, Epsilon Indi A, is a K star with one known planet, a very cold super-Jovian of about three Jupiter masses [17]. It has one of the longest periods known of any exoplanet. Not much more is known about it at present, although its proximity to our solar system, its large size and its decent separation from its star make it a likely target for telescope observation. The JWST has already had a look, although the results have only just been published [18].

A planet in the Epsilon Indi system is the setting of a novel based on the Halo universe. This remarkable series of combat-based video games has spawned books and a TV series. In the novel *Halo: Contact Harvest* (2007), the setting is the twenty-sixth century, and humans have established many interstellar colonies [19, 20]. There they encounter the Covenant, a religiously-inspired alien race bent on the destruction of humanity. The book tells the tale of the first contact between the two races, which as usual does not go well. The planet Harvest is described as a very fertile world despite being only about one-third the size of Earth. It is iceless and fairly warm, with a super-continent covering about two-thirds of its surface. The novel achieved the remarkable feat of entering the New York Times best seller list at #3.

Also remarkable about the Epsilon Indi system is that at least 1500 a.u. away from Epsilon Indi A is a pair of T dwarfs, orbiting each other with a separation of a couple of a.u. This is not the only binary pairing of T dwarfs in the known galaxy: the Luhman 16 system (Chap. 5) is even closer to Earth. For the Epsilon Indi brown dwarfs, the T1 dwarf is likely to be cloudier than the T6 one, and some variations in brightness of the pair have been tentatively ascribed to cloud variations in the atmosphere of the T1 dwarf [21]. These clouds can be composed of unusual materials. For instance, for the L4.5 brown dwarf MASSW J2224438-015852, in this sector at a distance of 11.6 parsecs, spectral observations combined with modeling suggest that its upper-level clouds are made of quartz, a crystalline form of silicon dioxide, and enstatite, a compound of magnesium, silicon and oxygen [22]. Lower down in the hotter, denser part of the atmosphere, there is a deck of iron clouds. Strange, but after the sapphire clouds of Dimidium (Chap. 3) and the forsterite clouds of Simpy (Chap. 7), maybe not so much.

A useful detour is to the home planet of the Atreides family of *Dune*, the ocean planet Caladan, circling the old G star Delta Pavonis in this sector. In the book, the Atreides family is "rewarded" by the Emperor with the rule of the important desert world Arrakis, the only place in the universe where the drug melange can be found. Melange is incredibly valuable, as it prolongs life and gives an ability to predict the future, as well as being essential for interstellar travel. Despite the obvious importance of this new assignment, the head of the Atreides family, Duke Leto, is suspicious of this offer, but feels duty-bound to accept it. Needless to say, things do not go smoothly on Arrakis. Delta Pavonis is a Sun-like star and has typically been considered a strong candidate to host a habitable world. It is billions of years older than the Sun, though [23, 24]. This might count against it as a location for a habitable terrestrial world, with any life-bearing worlds being worn out by this age (like 61 Virginis—see Chap. 2). In any event, no confirmed planets have yet been found, although a Jovian on a wide orbit is suspected [25].

Still, we don't have to go as far as Arrakis to find a desert world. There is one right next door: Mars.

Further Afield

While the presence or absence of an atmosphere is still an open question for most exoplanets, there is one in this sector where the answer is more clear-cut. The super-Earth planet Kua'kua (LHS 3844 b) orbits an M star in this sector about 15 parsecs away. In the Bribri language of Central America, Kua'kua means "butterfly", but given that the planet's day side temperature has been measured at about 1040 K, there is probably an absence of lepidoptera [26, 27]. Work published in 2019 by Laura Kreidberg of Harvard and her collaborators suggests that the temperature of the night side is "consistent with" absolute zero. Even though thermodynamically it actually has to be a little warmer than that, the night side is clearly a lot colder than the day side. This strongly implies a thin or absent atmosphere, one that is unable to transfer much heat from the day to the night side. The observations are a good fit to bare rock, so this could very well be a planet without a significant atmosphere. The absence of an atmosphere may be caused by a lack of volatiles in the planet's mantle, thus leading to weak outgassing. This could happen if the planet formed in the warmer part of the stellar system where volatiles were driven off by stellar radiation.

A similar planet in this sector is the Earth-sized GJ 1252 b, which has an even hotter day side. In 2022, Ian Crossfield of the University of Kansas and his collaborators made the point that even though their observations could only say that the surface atmosphere pressure of this planet was less than 10 times that of Earth, calculations suggest that any atmosphere thinner than 100 times that of Earth would have been wiped out by stellar radiation in less than 1 million years [28]. Outgassing of carbon dioxide could replace the atmosphere, but only if some fairly unreasonable assumptions were made about the amount of carbon in the planet's mantle. Thus the most likely outcome is a very thin atmosphere or no atmosphere.

Some planets don't get a chance to survive, let alone develop a stable atmosphere. Take the stellar system of HD 172555, an A star in this sector about 29 parsecs away in the constellation Pavo. This system, about 20 million years old, is still in the throes of development. About 5 a.u. out from the star, there is fine dust with an unusual composition. Spectral observations suggest that it is composed of silicates, with a spectrum typical of rocks formed from extreme melting events. There is also some silicate gas, formed by the vaporization of silicate rocks. Putting this all together, in 2009 Carey Lisse of Johns Hopkins and collaborators concluded that this was evidence of a massive collision between two planetary-sized bodies, possibly only a few thousand years ago [29]. This was big news back then, but the march of science has a way of trumping itself. In 2023, astronomers actually observed such a collision more or less while it was happening [30]. In this case, the brightness of the collision suggested that the colliding planets were super-Earth size or larger. The star in question is very distant and nowhere near the known galaxy, but this event serves as a salient reminder of the random, violent nature of stellar system formation and evolution.

Further out in this sector about 90 parsecs away is the Southern Hemisphere's miserable excuse for a pole star, Sigma Octantis, a barely visible dot that is in reality a very substantial F sub-giant about 40 times as luminous as the Sun

[31]. Good news, though, for our Southern Hemisphere readers: in about the year 66,000 C.E., the southern pole star will be … drum roll please… Sirius! [32] Sirius has a very large proper motion and in only this relatively short period of time (astronomically speaking), it will be in a completely different part of the sky than it is now. Take that, Polaris.

We end our tour of the known galaxy at a special place.

Of all of the exoplanet systems discovered to date, arguably the most fascinating of them all is the system of TRAPPIST-1, about 12.5 parsecs away in Aquarius. Although the name suggests that it was discovered by a group of medieval monks, the acronym stands for the TRAnsiting Planets and PlanetesImals Small Telescope, based at the La Silla Observatory on a desert mountain top in Chile. Like the SPECULOOS telescopes, TRAPPIST refers to a Belgian food product, in this case a popular beer [33]. The star itself is a very dim M star, category M8, with a tiny luminosity less than a thousandth that of the Sun. It has seven known planets, and all of them transit, so their characteristics are known reasonably accurately (see Fig. 9.2 and Table 9.2). All of the planets are small, some comparable in size to Earth, and several are

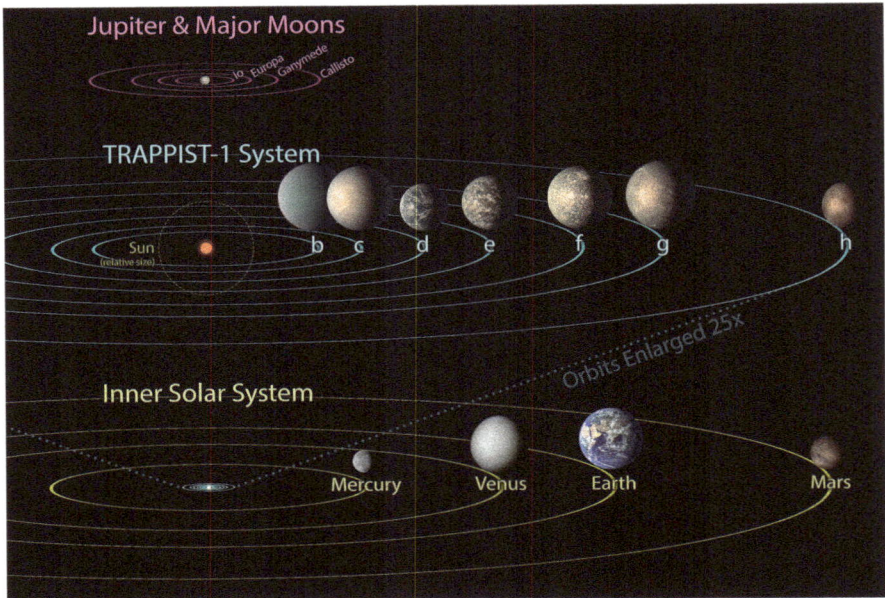

Fig. 9.2 The orbits of the planets of the TRAPPIST-1 system compared with the orbits of our solar system planets, along with the orbits of the moons of Jupiter. The orbits of the moons of Jupiter and of the TRAPPIST-1 planets have been enlarged by 25 times. Reprinted from https://photojournal.jpl.nasa.gov. © 2018, NASA/JPL-Caltech. Public domain

Table 9.2 Planets of TRAPPIST-1

Planet	Mass (Earth = 1)	Orbital period (days)	Distance from star (Earth distance from Sun = 1)	Equilibrium temperature (degrees K) (Earth = 255 K) (assumes zero albedo)	Type
TRAPPIST-1 b	1.37	1.5	0.011	405	Hot Earth-size
TRAPPIST-1 c	1.31	2.4	0.015	348	Hot Earth-size
TRAPPIST-1 d	0.39	4.0	0.021	295	Warm to hot rocky
TRAPPIST-1 e	0.69	6.1	0.028	255	Warm rocky
TRAPPIST-1 f	1.04	9.2	0.037	221	Warm Earth-size
TRAPPIST-1 g	1.32	12.4	0.045	201	Cold Earth-size
TRAPPIST-1 h	0.33	18.8	0.063	170	Cold rocky

within the conservative habitable zone. There has been a recent avalanche of scientific literature on these planets and wading through it all to arrive at a considered, up-to-date synthesis is challenging. Thus some conclusions stated here might very well be soon out of date.

But we do know that planet b is hot because its day side temperature has been measured by the JWST, finding a day side average of about 500 K [34]. This is very close to the theoretical value for a planet with zero albedo combined with no redistribution of heat to the night side, implying a thin or absent atmosphere. What might this atmosphere consist of? Based on theoretical studies, carbon dioxide is likely to be a component of the atmosphere of hot, small planets like planet b. The JWST measurements imply that if the atmosphere were composed entirely of carbon dioxide, its surface pressure would have to be considerably less than that of Mars [35]. If carbon dioxide were only one constituent amongst other gases, like nitrogen or water vapor for instance, the surface pressure would still most likely be less than 0.1 bar (10% of the surface pressure of Earth's atmosphere). An extended hydrogen atmosphere is also ruled out [36]. Some types of atmospheres cannot be ruled out yet, like a one bar nitrogen atmosphere with some methane, similar to Titan's atmosphere. The planet is very black, with a likely albedo of about 0.02, or darker than fresh asphalt. It will likely also experience a significant amount of tidal heating, about twice that of Io, because the eccentricity of its orbit is maintained through gravitational interactions with the other planets in the system [37]. Whether this large tidal heating would mean lots of surface volcanoes is not clear, because that would depend on the internal composition of the planet, and this is not certain at this time [38].

Planet c is slightly cooler but still located closer to its star than the runaway greenhouse limit. Measured temperature on the day side is 380 K. If a zero surface albedo is assumed, this temperature is between the calculated temperature assuming no heat redistribution by the atmosphere (i.e. bare rock) at 430 K and the value for a thick atmosphere capable of significant heat redistribution at 340 K [39, 40]. This suggests either some heat redistribution by an atmosphere or a higher albedo, or some combination of the two effects. Results also suggest that a thick atmosphere of 10 bars or more is unlikely, so a Venus-like atmosphere, either cloudy or cloud-free, is definitely excluded. A carbon-dioxide dominated atmosphere cannot be greater than 0.1 bar, while an 0.1 bar atmosphere of abiotic oxygen with some carbon dioxide is not yet ruled out. Simulations of planetary evolution using the PACMAN model also suggest the possibility of an abiotic oxygen atmosphere for planets b and c [41]. Also possible is an 0.1 bar steam atmosphere. That's a lot of possibilities, but one day soon we will know more.

Planet d initially looks more promising, with a much lower effective temperature of only about 295 K (about 22 °C; 72 °F). But planet d is small, bigger than Mars but smaller than Earth. SEPHI indicates a weak magnetic field, and this might mean significant atmospheric erosion in the past, despite the planet still having an outside chance of having liquid water on its surface. Climate modeling performed in 2018 by Andrew Lincowski of the University of Washington and collaborators did not give optimistic results, however [42]. Most simulated outcomes for this planet led to a runaway greenhouse. Some simulations created a dried-out, desert world with a 10–100 bar abiotic oxygen atmosphere, the relic of a long-lost ocean. For these simulations, habitable temperatures were produced, but this type of atmosphere may not occur very frequently in reality.

It is planet e that is the current focus of interest. Climate model simulations for a wide range of assumed atmospheres give habitable conditions, although assuming a Venus-like atmosphere creates a hot, uninhabitable planet [43, 44]. More recent PACMAN simulations have incorporated the latest observational results and now are constrained by the likely thin atmospheres of planets b and c, which reflect a smaller volatile inventory for those planets than assumed before [45]. To perform this new work, in 2023 Joshua Krissansen-Totton had to re-analyze PACMAN simulations published only in 2022, which gives you an idea of how fast our understanding of this system is developing. As it turns out, though, just because planet b might not have much of an atmosphere, for the vast majority of these new simulations planet e still retains a substantial atmosphere and the possibility of surface liquid water. This is largely because planet e is further away from its star than planet b, and so experiences less atmospheric loss. Another factor is that the planet has a slightly lower density than Earth does, and so may have a higher inventory of volatiles [38]. This might imply more water, for instance, and a terrestrial planet with a deep worldwide ocean is one possibility.

Available observations tell us little about the atmosphere of planet e. An extended hydrogen-helium atmosphere has been excluded, but for now that's about all that can be said [46]. If we make the massive but not implausible assumption that it has a 1 bar carbon-dioxide atmosphere, climate models show that for a dry planet, sub-stellar temperatures for this likely synchronously-rotating world are about 330 K, or uninhabitably hot, whereas night-side equatorial temperatures are about 200–230 K, or well below freezing [44]. For an aquaplanet, sub-stellar temperatures are lower, roughly 300–315 K (27–42 °C; 81–108 °F), but night-side temperatures are much higher, although still mostly below freezing [47]. If an Earth-like atmosphere is assumed instead, temperatures are lower and much of the planet is ice-covered, but there is still a large region of the day side with habitable temperatures. The upshot of all of this is that habitable conditions on planet e are not excluded at present.

If we do assume that the day side of the planet contains some habitable regions, what would surface conditions be like? The first difference from Earth would be the appearance of the planet's sun, TRAPPIST-1. The star would be about four times as wide in the sky of planet e as our Sun appears in our sky. Depending on the ability of the atmosphere of planet e to scatter radiation, TRAPPIST-1 would also appear redder, particularly in parts of the planet where it was low on the horizon. Surface gravity on planet e would be about 80% that of Earth [38]. Like many M stars, TRAPPIST-1 is a flare star, and experiences about four superflares per year [48]. A quick calculation shows that these would increase the brightness of the star very noticeably, but recent results suggest that these flares do not contain enough UV to seriously affect UV-resistant life forms on the surface of the planet, even if the planet had no ozone layer [49]. The other planets of the TRAPPIST-1 system would be easily visible from the surface of planet e. At closest approach, planet f would be as large in the sky of planet e as our Moon is in our sky, and like our Moon would display noticeable phases. The other planets would be smaller in the sky but even the most distant planet, planet h, would still be perceived as a small disk rather than a star. There would also be the usual meteorological conditions associated with synchronous rotation: persistent cloudiness on the day side and windy weather over much of the planet caused by the strong day-night temperature contrast.

Planet f is further away from TRAPPIST-1 and receives less radiation than does Mars, which is not a good omen for its habitability. It is a lot bigger than Mars, though, and so has more potential to retain an atmosphere. Its current water mass fraction is considerably higher than Earth's [50], and modeling suggests that the most likely scenario is that the planet is covered by a thick ocean with an atmosphere that may be rich in abiotically produced oxygen. A similar scenario has been proposed for planet g. Finally, planet h is the smallest, coldest and least dense of all of the TRAPPIST-1 worlds [38]. Observations rule out a clear primordial atmosphere of hydrogen and helium, and are consistent either with no atmosphere, or an atmosphere with a high cloud deck or haze layer [51].

Surface conditions for the three outermost TRAPPIST worlds (f, g and h) are still not certain at this time, and one possible scenario is that they are ocean worlds that are largely ice-covered. Assuming this, NASA modeling (Chap. 6) suggests ice thicknesses of several kilometers, with the ice shell sitting on top of an ocean [52]. The thinnest ice shell would be for the warmest world, planet f, and for that planet it might be possible for water geysers to vent particles into space. These particles might in future be detected by telescopes and would provide a strong indication of the presence of a liquid water ocean.

Since the TRAPPIST worlds are new, there is not a great deal of speculative fiction written about them yet. In the short story *The Eiffel Tower of TRAPPIST-1 d*, Jeff Reynolds describes a dying frontier world that comes up with a novel way to get back on its feet, taking advantage of some spectacular rock formations [53]. There is also *Trappist-1*, the third instalment of the Mark Noble space adventure series [54]. Here, our intrepid NASA astronauts discover that while planet e is cooler than Earth, it has liquid water and alien life, including leaping beluga-like sea creatures with dangerous tusks. As usual in these tales, the planet also hides a dark secret. The astronauts find that Planet f is no more welcoming, as it is colder and just as hostile, but for different reasons. No doubt there will be other stories told about these worlds, particularly when we know more about them.

Fomalhaut Sector Summary

Airless planets, dusty disks, colliding worlds: there are plenty of hostile places in this sector. But there may be some that are more welcoming. The TRAPPIST-1 system is crying out for further exploration. But it is just one of many systems in Aquarius, and one of many others that will soon be explored in the known galaxy. This new era of exoplanet observations and analysis might not exactly be the Age of Aquarius, but it is exciting enough. In the next chapter, we examine some of the big questions that exoplanet exploration may help to answer.

References

1. Mamajek EE (2012) On the age and binarity of Fomalhaut. Astrophys J Lett 754(2):L20. https://doi.org/10.1088/2041-8205/754/2/L20
2. Gáspár A, Wolff SG et al (2023) Spatially resolved imaging of the inner Fomalhaut disk using JWST/MIRI. Nat Astronom 7(7):790–798. https://doi.org/10.1038/s41550-023-01962-6
3. Gáspár A, Rieke GH (2020) New HST data and modeling reveal a massive planetesimal collision around Fomalhaut. Proc Natl Acad Sci USA 117(18):9712–9722
4. Le Guin UK (1964) The dowry of Angyar. In: Amazing Stories, September 1964, pp 46–63
5. Le Guin UK (1966) Rocannon's world. Ace, New York
6. Wikipedia (2024) Aquarius. Accessed 30 Mar 2024
7. Millholland S, Laughlin G, Teske J et al (2018) New constraints on Gliese 876—exemplar of mean-motion resonance. Astronom J 155(3):106
8. Jackson B, Greenberg R, Barnes R (2008) Tidal heating of extrasolar planets. Astrophys J 681(2):1631
9. Barnes R, Mullins K, Goldblatt C et al (2013) Tidal Venuses: triggering a climate catastrophe via tidal heating. Astrobiology 13(3):225–250
10. Martinez-Rodriguez H, Caballero JA, Cifuentes C et al (2019) Exomoons in the habitable zones of M dwarfs. Astrophys J 887(2):261
11. Correia ACM, Couetdic J, Laskar J et al (2010) The HARPS search for southern extra-solar planets-XIX. Characterization and dynamics of the GJ 876 planetary system. Astronom Astrophys 511:A21
12. Donati J-F, Cristofari PI, Finociety B et al (2023) The magnetic field and multiple planets of the young dwarf AU Mic. Mon Not Roy Astronom Soc 525(1):455–475. https://doi.org/10.1093/mnras/stad1193
13. Rockcliffe KE, Newton ER, Youngblood A et al (2023) The variable detection of atmospheric escape around the young, hot Neptune AU Mic b. Astronom J 166(2):77
14. Owen JE, Lai D (2018) Photoevaporation and high-eccentricity migration created the sub-Jovian desert. Mon Not Roy Astronom Soc 479(4):5012–5021. https://doi.org/10.1093/mnras/sty1760
15. Feinstein AD, France K, Youngblood A et al (2022) AU Microscopii in the far-UV: observations in quiescence, during flares, and implications for AU Mic b and c. Astronom J 164(3):110
16. Wittrock JM, Plavchan PP, Cale BL et al (2023) Validating AU Microscopii d with transit timing variations. Astronom J 166(6):232

17. Matthews EC, Carter AL, Pathak P et al (2024) A temperate super-Jupiter imaged with JWST in the mid-infrared. Nature 633:789–792

18. Space Telescope Science Institute (2024) Program information. https://www. stsci.edu/jwst/science-execution/program-information?id=2243.   Accessed 31 Mar 2024

19. Halopedia (2024) Halo. https://www.halopedia.org/Halo_universe. Accessed 1 Apr 2024

20. Staten J (2007) Halo: contact harvest. Tor, New York

21. Hitchcock JA, Helling C, Scholz A et al (2020) Large-scale changes of the cloud coverage in the ϵ Indi Ba and Bb system. Mon Not Roy Astronom Soc 495(4):3881–3899. https://doi.org/10.1093/mnras/staa1344

22. Burningham B, Faherty JK, Gonzales EC et al (2021) Cloud busting: enstatite and quartz clouds in the atmosphere of 2M2224-0158. Mon Not Roy Astronom   Soc   506(2):1944–1961.   https://doi.org/10.1093/mnras/stab1361

23. Porto de Mello GF, del Peloso EF, Ghezzi L (2006) Astrobiologically interesting stars within 10 parsecs of the Sun. Astrobiology 6(2):308–331. https://doi.org/10.1089/ast.2006.6.308

24. Mamajek EE, Hillenbrand LA (2008) Improved age estimation for solar-type dwarfs using activity-rotation diagnostics. Astrophys J 687(2):1264–1293. https://doi.org/10.1086/591785

25. Makarov VV et al (2021) Looking for astrometric signals below 20 m/s: a candidate exo-Jupiter in δ Pav. Res Not Am Astronom Soc 5(5):108. https://doi.org/10.3847/2515-5172/abfec9

26. IAU (2024) 2022 approved names. https://www.nameexoworlds.iau.org/2022approved-names. Accessed 2 Apr 2024

27. Kreidberg L, Koll DDB, Morley C et al (2022) Absence of a thick atmosphere on the terrestrial exoplanet LHS 3844 b. Nature 573:87–90. https://doi.org/10.1038/s41586-019-1497-4

28. Crossfield IJ, Malik M, Hill ML et al (2022) GJ 1252 b: a hot terrestrial super-Earth with no atmosphere. Astrophys J Lett 937(1):L17

29. Lisse CM, Chen CH, Wyatt MC et al (2009) Abundant circumstellar silica dust and SiO gas created by a giant hypervelocity collision in the ~12 Myr HD 172555 system. Astrophys J 701(2):2019

30. Kenworthy M, Lock S, Kennedy G et al (2023) A planetary collision afterglow and transit of the resultant debris cloud. Nature 622(7982):251–254

31. Houk N (1975) Michigan catalogue of two-dimensional spectral types for the HD stars. University of Michigan, Ann Arbor, MI

32. McClure B (2023) Sirius is a future southern Pole Star. earthsky.org. https://earthsky.org/tonight/sirius-future-south-pole-star/. Accessed 2 Apr 2024

33. Wikipedia (2024) Trappist beer. Accessed 3 Apr 2024

34. Greene TP, Bell TJ, Ducrot E et al (2023) Thermal emission from the Earth-sized exoplanet TRAPPIST-1 b using JWST. Nature 618:39–42. https://doi.org/10.1038/s41586-023-05951-7

35. Ih J, Kempton EMR, Whittaker EA, Lessard M (2023) Constraining the thickness of TRAPPIST-1 b's atmosphere from its JWST secondary eclipse observation at 15 μm. Astrophys J Lett 952(1):L4

36. Lim O, Benneke B, Doyon R et al (2023) Atmospheric reconnaissance of TRAPPIST-1 b with jwst/niriss: Evidence for strong stellar contamination in the transmission spectra. Astrophys J Lett 955(1):L22

37. Barr AC, Dobos V, Kiss LL (2018) Interior structures and tidal heating in the TRAPPIST-1 planets. Astronom Astrophys 613:A37

38. Agol E, Dorn C, Grimm SL et al (2021) Refining the transit-timing and photometric analysis of TRAPPIST-1: masses, radii, densities, dynamics, and ephemerides. Plan Sci J 2(1):1

39. Zieba S, Kreidberg L, Ducrot E et al (2023) No thick carbon dioxide atmosphere on the rocky exoplanet TRAPPIST-1c. Nature 620:746–749. https://doi.org/10.1038/s41586-023-06232-z

40. Lincowski AP, Meadows VS, Zieba S et al (2023) Potential atmospheric compositions of TRAPPIST-1 c constrained by JWST/MIRI observations at 15 μm. Astrophys J Lett 955(1):L7

41. Krissansen-Totton J, Fortney JJ (2022) Predictions for observable atmospheres of TRAPPIST-1 planets from a fully coupled atmosphere–interior evolution model. Astrophys J 933(1):115

42. Lincowski AP, Meadows VS, Crisp D (2018) Evolved climates and observational discriminants for the TRAPPIST-1 planetary system. Astrophys J 867(1):76

43. Ibid.

44. Turbet M, Fauchez TJ, Sergeev DE et al (2022) The TRAPPIST-1 Habitable Atmosphere Intercomparison (THAI). I. Dry cases—the fellowship of the GCMs. Plan Sci J 3(9):211

45. Krissansen-Totton J (2023) Implications of atmospheric nondetections for TRAPPIST-1 inner planets on atmospheric retention prospects for outer planets. Astrophys J Lett 951(2):L39

46. de Wi J, Wakeford HR, Lewis NK et al (2018) Atmospheric reconnaissance of the habitable-zone Earth-sized planets orbiting TRAPPIST-1. Nat Astronom 2:214–219. https://doi.org/10.1038/s41550-017-0374-z

47. Sergeev DE, Fauchez TJ, Turbet M et al (2022) The TRAPPIST-1 Habitable Atmosphere Intercomparison (THAI). II. Moist cases—the two waterworlds. Plan Sci J 3(9):212

48. Glazier AL, Howard WS, Corbett H et al (2020) Evryscope and K2 constraints on TRAPPIST-1 superflare occurrence and planetary habitability. Astrophys J 900(1):27

49. Estrela R, Palit S, Valio A (2020) Surface and oceanic habitability of TRAPPIST-1 planets under the impact of flares. Astrobiology 20(12):1465–1475

50. Barth P, Carone L, Barnes R et al (2021) Magma ocean evolution of the TRAPPIST-1 planets. Astrobiology 21(11):1325–1349. https://doi.org/10.1089/ast.2020.2277
51. Gressier A, Mori M, Changeat Q et al (2022) Near-infrared transmission spectrum of TRAPPIST-1 h using Hubble WFC3 G141 observations. Astronom Astrophys 658:A133
52. Quick LC, Roberge A, Mendoza GT et al (2023) Prospects for cryovolcanic activity on cold ocean planets. Astrophys J 956(1):29
53. Reynolds J (2023) The Eiffel Tower of TRAPPIST-1 d. In: Analog science fiction and fact, November/December 2023, pp 38–50
54. Harmsworth T (2020) Trappist-1 (Mark Noble Adventure #3). www.harmsworth.net

10

The Unknown Known Galaxy

Our knowledge of the known galaxy is in a tremendous state of flux. There is a flood of new information, and today's scientific consensus may soon turn out to be yesterday's news. This field is moving so rapidly that some of the material presented here may be out of date in less than 10 years. Even so, much more is now understood about the known galaxy than at any time in history. As we have learned more, a number of the major assumptions made by scientists and science fiction writers in past decades have been challenged by new science. This chapter will outline some of those topics and discuss what still needs to be learned about this crucially important part of the galaxy. We will also address the issue of whether we are starting to establish a sense of place for this region, and how this might eventuate.

Perhaps most important of these changing assumptions has been the contradiction of the idea that the solar system is typical. Our solar system is weird. Scientists expected to find that most stellar systems would conform to the solar system model of smaller rocky planets close to the star and bigger gassy planets further out. That was not what was found. Instead, we discovered that the solar system is atypical. In the known galaxy and elsewhere, many gassy planets were found orbiting close to their stars. Few gassy planets were found on wide orbits, like Jupiter and Saturn. Systems were found that might have only a couple of planets, or even just one big one with a very eccentric orbit. Many systems have a whole bunch of planets crammed into orbits closer than Mercury's distance from the Sun. The solar system combination of Earth-size or smaller planets in the inner solar system with large ones in the outer system appears to be rare, at less than 1% of systems [1].

K. J. E. Walsh, *Planets of the Known Galaxy*, Science and Fiction,
https://doi.org/10.1007/978-3-031-68218-6_10

More common are systems with inner planets that are larger than the inner planets of our solar system. In our solar system, this situation may not have arisen because Jupiter's gravity acted to remove material from the inner solar system, thereby decreasing its volatile content. This created a solar system like ours where the inner planets were smaller and depleted in water, as opposed to many other systems where the inner planets are super-Earth or sub-Neptune ocean worlds [2, 3]. In fact, it has been speculated that a giant planet in our outer solar system was actually needed to create an Earth-like world with modest rather than overwhelming amounts of water. This idea might explain why we are living in a stellar system that is in a definite minority, rather than in a typical system as was once supposed. We need to be atypical in order to exist.

These new discoveries have left scientists scrambling for explanations and science fiction writers exploring a whole new playground. Now that we know some of the specific details of a few stellar systems, this knowledge is gradually being incorporated into speculative fiction.

Habitable planets might exist around M stars. Before the modern era of exoplanet exploration, it was expected that there might be an occasional world circling an M star that could be habitable. But such planets might be more abundant than previously thought. Climate simulations have suggested that a planet that keeps the same face towards its star is not necessarily a blazing desert on one side and a frozen wasteland on the other. Heat transport by the ocean and atmosphere from the day side to the night side can be effective in reducing temperature contrasts. Persistent cloud on the day side can lower the temperature there as well. As a result, these tidally locked planets still have the potential to be habitable. Moreover, it has been shown that planets with thick atmospheres might be able to avoid becoming locked into synchronous rotation [4]. A slowly rotating but non-synchronous planet might be more habitable in some ways than a synchronously rotating one, as more regions of the planet would periodically experience warmer temperatures.

True, there are important limitations for habitability of planets orbiting M stars. Worlds larger than Earth circling these stars might have global oceans but no land surface [5, 6]. There are a number of reasons for this. The dust disks of M stars are smaller, denser, and contain more volatiles, and this would mean more water. M-star systems do not have many giant planets, and so the gravity of these larger worlds cannot clear out the inner part of the system. The snow line is closer to the habitable zone in M stars than in G stars, thus making it easier for habitable zone planets to collect volatiles. Thus many Earth-size M star planets in the habitable zone are likely to be waterworlds. Planets with deep oceans would have a high-pressure ice layer at the bottom of the sea, potentially suppressing volcanism and cutting off the climate stabilizing effect of the carbonate-silicate cycle (Fig. 2.4). These planets might suffer from strong climate variability, leading to either runaway greenhouse or a snowball state. Having said this, our understanding of this process is preliminary at this stage, to say the least.

Both the oceans and atmospheres of planets circling young M stars might be blasted away by flares, and a desert world might be created. While these dry worlds would be more resistant to runaway greenhouse, and would also be less likely to freeze, it would be a delicate balancing act to create a habitable desert Earth-size world. Too little removal of atmosphere and the planet would still be an ocean world. Too much removal and a Mars- or Moon-like planet would be created, largely hostile to life. The just-right desert world with some water but not too much might be rare.

Some planets around M stars might have tidal heating rates that make them too volcanic, even to the point of having molten surfaces. For their habitability to be destroyed, they would not even have to be experiencing these high tidal heating rates today. If they were strongly tidally heated in their youth, that might have been enough to push them into a permanent runaway greenhouse and create a tidal Venus.

At present, there are different opinions on the likelihood that M stars will host a lot of Earth-like planets. Having said this, there are just so many M stars that worlds circling them may form an important class of habitable planets.

While they are not very hospitable to human beings, oceans beneath frozen surfaces might host life more frequently than terrestrial planets. In our own solar system, there are many smaller worlds that probably have large oceans beneath frozen surfaces, like Europa and Enceladus, so there are likely to be many in other solar systems as well. There may also be some travelling between the stars, like the Steppenwolves and other rogue planets of various sizes. In our solar system, Europa is a prime target of NASA's Europa Clipper mission [7]. Scheduled for launch in 2024, its main task is to find out whether Europa's ocean could support life. To do this, it will determine the thickness of the ice shell and confirm whether there really is any water underneath it. It will investigate the composition of the ocean, and it will see whether there is any recent geological activity, such as geysers, that might deposit ocean water on the surface. Later missions, still in the concept stage, envisage robot spacecraft landing on Europa or even drilling through its ice shell and sampling its ocean [8]. Manned missions would be dangerous because Europa is within Jupiter's giant radiation belts and an unprotected astronaut standing on its surface for 1 day would receive a lethal dose of radiation [9]. On the other hand, one of

Jupiter's other moons, Callisto, is outside Jupiter's belts and receives much less radiation, so it could serve as a manned base for the exploration of the rest of the system. The search for life on Europa is active and well backed by science. If life is found under the ice of Europa, this greatly increases the chances of life on other Europa-like worlds in the known galaxy.

While life in such worlds could be plentiful, there are many ways that the evolution of a planet can go wrong. Runaway greenhouse, having too much or too little water, being tossed out of a stellar system altogether; these are some of the ways that potentially habitable planets end up being uninhabitable. Runaway greenhouse is real, since we have a convincing example in our own solar system, the unlucky world we call Venus. But while runaway greenhouse is well established in theory, the precise conditions under which it can occur on a planet remain uncertain. There is a lack of observations to constrain precisely the inner limits of the habitable zone. For dry or slowly rotating worlds, calculations suggest that runaway greenhouse is less likely, but again there are no observations to validate these predictions.

While dry worlds might be able to maintain habitable temperatures in regions inside the conservative habitable zone, by definition they would have little water. Life as we know it needs water. There may well be life as we do not know it, perhaps using ammonia as a solvent instead of water [10]. But ammonia is a liquid only at temperatures less than −33 °C (−27 °F) at the atmospheric pressure at terrestrial sea level. Ammonia-based life would inhabit locations that are cold and very inhospitable to humans.

Too much water is a problem as well. A planet with no land would have limited habitability from a human perspective. As already discussed, a planet with a deep ocean might experience high climate variability, making a runaway greenhouse or snowball state more likely [11]. While there do seem to be some common factors that influence the final water content of a planet, simulations indicate that the amount of water that a terrestrial planet ends up with seems to be very random.

Randomness can also affect whether a planet even stays in a stellar system. Planets can be tossed out of stellar systems by gravitational interactions with other planets. Some binary star systems form with their stars located too close to each other to permit the formation of habitable zone planets. Stellar systems have been discovered with giant planets on very eccentric orbits that would have spelled doom for any habitable planets. Even in relatively well-behaved places like our own solar system, planets are thought to have migrated around during the early evolution of their orbits, throwing at least one planet out into interstellar space, the planet that we have called Chaos. This world might still be hanging around in the form of Planet Nine, but no one really knows where it is, and we may never know.

Even if a planet avoids being tossed out in the darkness between the stars, planets could just get old and tired. There are some perfectly reasonable stars out there that may be just too ancient to have planets with robust, functioning ecosystems. Planets that are too young have different problems. It takes a while for the remaining loose rocks floating around a stellar system to stop crashing into planets at unacceptably high rates, and in our solar system, it was not until after the main part of the Late Heavy Bombardment ended about 3.5 billion years ago that impact rates from incoming rocks started to settle down to reasonable levels [12]. Some systems have a lot more asteroids than our solar system, so their current impact rates may still be high, with potentially disastrous effects on the evolution of life on their habitable zone planets. Moreover, life on Earth was slow to develop to the point where it was able to create an atmosphere that was breathable by humans. Our own atmosphere has only been breathable in about the last 500 million years or so, only about one-tenth of the age of the planet [13]. Maybe life on other planets would develop faster, or slower—or maybe not at all. We just don't know.

Even though on our planet the spread of abundant life was the crucial factor in the development of a breathable atmosphere, on other worlds it is possible that this could have taken place without any life at all. This realization has been one of the positives of our exploration of the known galaxy, although such worlds are yet to be confirmed by observations. While the process that causes the non-biological generation of an oxygen-rich atmosphere, photodissociation of water, would be most likely on hot, wet worlds, simulations show that it is also possible for it to occur on drier, more Earth-like planets. The downside of this is that the detection of an oxygen-rich atmosphere on a planet is no longer regarded as a completely unambiguous sign of life.

There are likely other environments where carbon-based life could exist but that are hostile to humans. A world in the outer part of the habitable zone may have hospitable temperatures but only because of a large and toxic amount in its atmosphere of the greenhouse gas carbon dioxide. A water world may have an ocean that is hotter than the terrestrial sea level boiling point, kept liquid only by a high-pressure atmosphere. Heat-loving organisms in this steaming ocean might be perfectly happy, but humans would not. Hycean worlds may have thick hydrogen atmospheres whose large greenhouse effect might be able to maintain a liquid ocean even in outer regions of a stellar system, but at the cost of a crushingly high surface atmospheric pressure.

While atmospheric differences will be important in determining planetary climates, the surfaces and compositions of many terrestrial planets will differ from Earth's. Some worlds will be very dense, the so-called "Super Mercuries", with a composition dominated by iron [14]. Other worlds, instead of having a crust that is mostly compounds of silicon like Earth's, might have a crust dominated by aluminium compounds, like Sapphire World. Or there may be carbon worlds, like (possibly) Janssen, with tar oceans and diamond lava.

Even though there are many planets that are not very Earth-like, extrapolation of the Kepler data suggests that about 24% of Sun-like stars should have an Earth-size planet in the habitable zone [15]. Other estimates suggest the percentages could be as high as 60%, and the nearest one around a G or K star could be only about 6 parsecs away. The same estimates suggest that there could be four such worlds inside the 10 parsec boundary chosen here for the known galaxy [16]. As of 2024, none have been found, although a number of Earth-sized worlds have been found circling M stars in their habitable zones.

Assuming then that Earth-size planets in the habitable zones of Sun-like stars do exist, how many of them will be habitable? Here the projections become extremely uncertain. Just because a planet orbits in the habitable zone does not mean that it is habitable, and an unknown percentage of such worlds will be either too hot or too cold. Of those planets that are temperate and have oceans, it is not clear how many will be life-bearing. And it is very unclear how many of those worlds will have enough life over a long enough period of time to produce an atmosphere with a substantial amount of oxygen in it. We simply do not have the data to make reasonable estimates of these factors yet.

This has not stopped scientists from making some informed speculation on the topic, however [17–19]. One of the reasons that scientists are so keen to find out if there is or ever has been life on Mars is that if it is discovered there, this would greatly increase the chance that there is life elsewhere in the known galaxy. If life is then found on a third or fourth world in our system, like Enceladus for instance, the working assumption would become that life would develop wherever the environment was hospitable for it to do so. The implication would be that life was common in the known galaxy. The crucial question would then become whether life would evolve in such a way as to create an oxygen atmosphere on an Earth-size planet. I suspect that telescopic observations will soon leapfrog this question: that is, we will probably know from observations how frequent are Earth-like, oxygen-bearing worlds around other stars before we have reliable estimates from theory. During the golden age of exoplanetary discovery, theory has often been playing catch-up, and data has been key.

The lack of data on Earth-like worlds is a substantial gap in our knowledge and will likely be addressed in the near future by new telescopes. The Nancy Grace Roman Space Telescope will be the successor to the JWST. It is named after Nancy Roman, who in 1959 was appointed NASA's first chief of astronomy [20]. In that year, she proposed a method to detect extrasolar planets that involved masking the light of the star to detect better the much dimmer light of a planet. This method was later used by the Hubble Space Telescope to take a picture of the potential planetary candidate orbiting Fomalhaut (Chap. 9). It will also be used on the Roman Space Telescope when it is launched, most likely in 2027. This telescope will be able to directly image Jovian worlds, as well as performing transit and radial velocity measurements. It will also do microlensing, and one project is to find out how many rogue planets there really are, down to about the size of Mars [21]. Scheduled for launch in 2026 is ESA's PLAnetary Transits and Oscillations of stars (PLATO) space observatory. It will concentrate on finding and characterizing Earth-size planets orbiting Sun-like stars, including those in the habitable zone [22]. A more advanced space-based telescope is on the drawing board. The Habitable Worlds Observatory will be able to image Earth-like worlds around nearby stars, and then analyze their atmospheres [23]. The race to find the first candidate Earth-like world is on.

While space-based telescopes have much better observing conditions than Earth-based telescopes, ground-based instruments have the huge advantage that they can be repaired and upgraded much more easily. The 39-m-wide, inventively named Extremely Large Telescope (ELT) is being constructed on a mountain in the Atacama Desert of northern Chile (perhaps the project managers need a better acronym…) [24]. When ready to start work in 2028, it will try to directly image Earth twins orbiting nearby stars and analyze their atmospheres. The ELT web site gives the impression that this will be very challenging for this telescope, however. Nevertheless, there will surely be bigger Earth-based telescopes in the future, with even greater capabilities.

The ultimate goal, of course, is to find evidence of life on Earth-like planets circling other stars. Which stars in particular? A list of the most promising candidates was compiled in 2023 by Eric Mamajek of NASA's Jet Propulsion Laboratory [25, 26]. The idea was to identify those stars that would have the most easily observable planets in their habitable zones. These planets would need to have a decent separation from their host stars to be observable, so while some of the F stars on the list are as far away as 24 parsecs, the M stars are limited to only 4 parsecs. This is due to the habitable zones of M stars being so much closer to them, making the light of any planets harder to separate from the light of the star. Many of the stellar suspects identified in this book are on the list, but there are others on the list that have not been mentioned in the previous chapters because to date no planets have been discovered in their stellar systems. Among the most interesting within our 10 parsec distance limit is the K0 star 40 Eridani A, part of a triple system that includes a white dwarf. This does not appear to be a deal-breaker, as the white dwarf is about 400 a.u. away from the K star, too far away to be a problem. Not to be outdone, the Gliese 570 system has four stars of classes K4, M1, M3, and a T7 brown dwarf. Again, though, the K4 star is quite distant from the other stars, and so it is on the list. The 36 Ophiuchi system has not one but three K stars, and all three of them are on the list, so this surely must be a high-priority target. The home system of Larry Niven's feline interstellar conquerors the Kzinti, the single G8 star 61 Ursa Majoris, is included. Even 82 Eridani is listed, despite the current uncertainty about the orbits of its detected planets (see Chap. 7). So are 61 Virginis, Delta Pavonis and Beta Hydri, despite their advanced ages. Naturally, Alpha Centauri A and B are there, but not Proxima, presumably because it is too dim. Even pi Mensae puts in an appearance, despite the star possessing a super-Jovian planet with an eccentric orbit that would surely interfere with the orbit of a habitable planet (see Chap. 8). Nevertheless, astronomers are prioritizing their searches so that when the time comes for them to use the precious observing time of the new generation of space telescopes, they will be ready.

To determine whether an exoplanet might have life, its atmospheric spectrum needs to be analysed for so-called biosignatures, the presence of particular molecules in its atmosphere that might indicate the presence of life. Life as we know it is based on carbon molecules, although it is not impossible that other biochemistries might exist elsewhere [27]. Water is also likely to be very important. There are many proposed biosignatures, so only a few will be mentioned here. For instance, methane and oxygen react with each other, so the only way for them to be present in an atmosphere together would be if they were continually being replenished somehow—by life, for instance. The JWST can detect carbon dioxide and water vapor on planets like those in the TRAPPIST-1 system, but not Earth-like levels of oxygen. Large amounts of abiotic oxygen (>1 bar) are another story, and are potentially detectable [28].

A new road map for the detection of biosignatures was recently outlined by Amaury Triaud of the University of Birmingham and co-authors [29]. They point out that while the JWST really cannot detect Earth-like levels of oxygen in exoplanet atmospheres, the ELT might be able to. But there is another method, one that the JWST could use. A planet that is life-bearing will likely have less atmospheric carbon dioxide, as it will have been absorbed by the biosphere and the hydrosphere (oceans). They point out that Earth is remarkable in the solar system for its low percentage of atmospheric carbon dioxide. The JWST can observe this quantity for terrestrial planets circling late M stars e.g. the TRAPPIST-1 system planets. If ozone and methane are also detected, these would be strong indicators of the presence of a biosphere. There are also biosignatures that might be measured on the surfaces of exoplanets. Light reflected from terrestrial vegetation has a particular spectrum, the "red edge". This signature might be realistically detectable by instruments flown on future space-based missions [30].

There is so much more that could be said about how we might be able to detect evidence of life around planets in the known galaxy and elsewhere. We are at the beginning of a golden age of exoplanet exploration, almost analogous to the golden age of planetary exploration that began in the 1960s and arguably is still under way. Robot space probes travelling through our solar system have given us pictures of worlds that were once just dots in the sky.

So what have we learned about some of the places in the known galaxy? A sense of place might include elements that are known to be fictional. For instance, the *Lord of the Rings* films changed our perception of New Zealand forever [31]. We know that Tolkien's fictional town of Hobbiton is not really located in the North Island countryside about two hours south of Auckland, but then again, maybe it is, just a bit.

The known galaxy is full of both fact and fiction. Proxima Centauri b is Chiron's Waterworld, a coolish super-Earth that could have a global ocean, with part of it frozen. It is likely windy and gloomy. Some 13,000 a.u. away is Alpha Centauri A, orbited by the large planet Polyphemus, whose habitable moon Pandora is the only source of the crucial mineral unobtanium. In the same region of space is Wolfy, either a real Venus of runaway greenhouse, or a tropical ocean world of old Venus. In the location of Maui's fishhook is the hot super-Earth GJ 667C c, or Hook c as we call it, which could be either a hot, watery world with a thick atmosphere, or a nearby example of a super-Venus, like Venus but bigger and even more hostile.

I could go on, but that would just be restating text from the previous chapters. Words like "could be" and "potentially" are used a lot in the previous paragraph. At our current level of understanding of this topic, this is unavoidable. But for extrasolar planets, there is a more serious issue: how does one establish a sense of place for a location that no one may ever visit?

Our neighboring desert planet Mars does not have this problem. Figure 10.1 shows a magnificent panorama of a small Martian valley, taken by the Curiosity rover. The landscape seems familiar to an Earthling, but only a little. It looks like an extremely arid part of Earth, but the rocks are darker and the sky is a pink-orange color. There is something definitely alien about this landscape— not surprising, because it is. Images like this are crucially important for giving this alien world a sense of place. The images that have been obtained from the surfaces of other worlds in our solar system—so far, the Moon, Venus, Titan

Fig. 10.1 A valley at the foot of Mount Sharp, Mars. The rock formation on the left is Mont Mercou and is about seven meters tall. Image taken by the Curiosity rover on March 5, 2021. Reprinted from www.planetary.org. © 2021, NASA/JPL-Caltech/MSSS/ Thomas Appéré. Public domain

and some asteroids—are also of landscapes that are recognizably non-terrestrial, and not just because of an absence of vegetation.

But we don't have to wait until we have put a rover with a camera on, say, the surface of Proxima Centauri b before we can start to get a sense of place for this world. Nor does this sense of place actually have to be entirely accurate. The perception of a place can change over time. The Mars of Percival Lowell's canals, of Barsoom, of Kim Stanley Robinson's brave terraformers in his Mars trilogy [32], of Ridley Scott's *The Martian*: these are all very different places, but they all serve to create a mental and emotional connection to the planet Mars that is at least partly based on reality.

This emotional connection, this sense of place, does not have to be completely real. It just has to be believed, if only a little. And so it is with the other planets of the known galaxy. They are far away, and because of the enormous and well-known difficulty of interstellar travel, many of these planets may never be visited by us. But as we know more about them, they will become more real. They will start to become places.

References

1. Zhu W, Dong S (2021) Exoplanet statistics and theoretical implications. Annu Rev Astronom Astrophys 59:291–336

2. Lambrechts M, Morbidelli A, Jacobson SA et al (2019) Formation of planetary systems by pebble accretion and migration—how the radial pebble flux determines a terrestrial-planet or super-Earth growth mode. Astronom Astrophys 627:A83

3. Kokaia G, Davies MB, Mustill AJ (2020) Resilient habitability of nearby exoplanet systems. Mon Not Roy Astronom Soc 492(1):352–368. https://doi.org/10.1093/mnras/stz3408

4. Leconte J, Wu H, Menou K, Murray N (2015) Asynchronous rotation of Earth-mass planets in the habitable zone of lower-mass stars. Science 347:632–635

5. Shields AL, Ballard S, Johnson JA (2016) The habitability of planets orbiting M-dwarf stars. Phys Rep 663:1–38

6. Ramirez RM (2018) A more comprehensive habitable zone for finding life on other planets. Geosciences 8(8):280

7. NASA (2024) Europa Clipper. https://europa.nasa.gov/. Accessed 7 Mar 2024

8. Weiss P, Yung KL, Kömle N et al (2011) Thermal drill sampling system onboard high-velocity impactors for exploring the subsurface of Europa. Adv Space Res 48(4):743. https://doi.org/10.1016/j.asr.2010.01.015

9. Zubrin R (2000) Entering space: creating a spacefaring civilization. Tarcher Perigee, New York, p 167

10. Rampelotto PH (2010) The search for life on other planets: Sulfur-based, silicon-based, ammonia-based life. J Cosmol 5:818–827

11. Dehant V, Debaille V, Dobos V et al (2019) Geoscience for understanding habitability in the solar system and beyond. Space Sci Rev 215:1–48

12. Bottke WF, Norman MD (2017) The late heavy bombardment. Annu Rev Earth Plan Sci 45:619–647

13. Holland HD (2006) The oxygenation of the atmosphere and oceans. Philos Trans Roy Soc B Biol Sci 361(1470):903–915. https://doi.org/10.1098/rstb.2006.1838

14. Mah J, Bitsch B (2023) Forming super-Mercuries: Role of stellar abundances. Astronom Astrophys 673:A17

15. Checlair JH, Villanueva GL, Hayworth BP et al (2021) Probing the capability of future direct-imaging missions to spectrally constrain the frequency of Earth-like planets. Astronom J 161(3):150

16. Bryson S, Kunimoto M, Kopparapu RK et al (2020) The occurrence of rocky habitable-zone planets around solar-like stars from Kepler data. Astronom J 161(1):36

17. Perryman M (2018) The exoplanet handbook, 2nd edn. Cambridge University Press, New York

18. Cockell CS, Bush T, Bryce C et al (2016) Habitability: a review. Astrobiology 16(1):89–117

19. Madau P (2023) Beyond the Drake equation: a time-dependent inventory of habitable planets and life-bearing worlds in the solar neighborhood. Astrophys J 957(2):66

20. Roman NG (2019) Nancy Grace Roman and the dawn of space astronomy. Annu Rev Astronom Astrophys 57:1–34. https://doi.org/10.1146/annurev-astro-091918-104446

21. NASA (2024) Nancy Grace Roman Space Telescope. https://roman.gsfc.nasa.gov/. Accessed 11 Mar 2024

22. ESA (2024) PLATO factsheet. https://www.esa.int/Science_Exploration/Space_Science/Plato_factsheet. Accessed 13 Mar 2024

23. Habitable Worlds Observatory (2024) Science. https://habitableworldsobservatory.org/science/. Accessed 11 Mar 2024

24. ESO (2024) Science with the ELT. https://elt.eso.org/science/exoplanets/. Accessed 11 Mar 2024

25. Mamajek E, Stapelfeldt K (2023) NASA ExEP mission star list of plausible targets for a future IROUV direct imaging space observatory. In: Abstracts of American Astronomical Society Meeting, Seattle, WA, 8–12 Jan 2023. https://ui.adsabs.harvard.edu/abs/2023AAS...24111607M/abstract

26. NASA Exoplanet Archive (2024) HWO ExEP precursor science stars. https://exoplanetarchive.ipac.caltech.edu/cgi-bin/TblView/nph-tblView?app=ExoTbls&config=DI_STARS_EXEP. Accessed 13 Mar 2024

27. Schwieterman EW, Kiang NY, Parenteau MN et al (2018) Exoplanet biosignatures: a review of remotely detectable signs of life. Astrobiology 18(6):663–708

28. Krissansen-Totton J, Fortney JJ (2022) Predictions for observable atmospheres of TRAPPIST-1 planets from a fully coupled atmosphere–interior evolution model. Astrophys J 933(1):115

29. Triaud AHMJ, de Wit J, Klein F et al (2024) Atmospheric carbon depletion as a tracer of water oceans and biomass on temperate terrestrial exoplanets. Nat Astron 8:17–29. https://doi.org/10.1038/s41550-023-02157-9

30. Barrientos JG, MacDonald RJ, Lewis NK, Kaltenegger L (2023) In search of the edge: A Bayesian exploration of the detectability of red edges in exoplanet reflection spectra. Astrophys J 946(2):96

31. The lord of the rings: the fellowship of the ring (2001) imbd.com. https://www.imdb.com/title/tt0120737/?ref_=nv_sr_srsg_3_tt_6_nm_2_q_the%2520lord%2520of%2520the%2520rin. Accessed 6 June 2024

32. Robinson KS (1993) Red Mars. Spectra, New York

www.ingramcontent.com/pod-product-compliance
Ingram Content Group UK Ltd.
Pitfield, Milton Keynes, MK11 3LW, UK
UKHW021437231224
452810UK00002B/23

9 783031 682179